Singapore Math 2.0

The Stack Model Method
An Intuitive and Creative Approach to Solving Word Problems

Grades 5–6

MATHPLUS Publishing

Yan Kow Cheong

MATHPLUS Publishing
Blk 639 Woodlands Ring Road
#02-35 Singapore 730639

E-mail: publisher@mathpluspublishing.com
Website: www.mathpluspublishing.com

First published in Singapore in 2016

Copyright © 2016 by Yan Kow Cheong

All rights reserved. No part of this publication may be reproduced, stored in a retrieval system, or transmitted in any form or by any means, electronic, mechanical, photocopying, recording or otherwise, without the prior written permission from the publisher.

National Library Board, Singapore Cataloguing-in-Publication Data

Yan, Kow Cheong, author.
 The stack model method, Grades 5-6 : an intuitive and creative approach to solving word problems / Yan Kow Cheong. – Singapore : MathPlus Publishing, 2015.
  pages cm
  ISBN : 978-981-09-3835-2 (paperback)

 1. Problem solving – Study and teaching (Primary). 2. Mathematics – Study and teaching (Primary). 3. Mathematics – Problems, exercises, etc. I. Title.

QA135.5
372.7044 -- dc23   OCN 903489185

Printed in the United States of America

Preface

You are familiar with the *Singapore Model Method*, or *Bar Method*, as a powerful visualization problem-solving strategy (or "problem-solving heuristic," as math teachers call it in Singapore). What about the more intuitive and insightful Singapore's *Stack Model Method* for solving mathematics word problems?

No longer do you need to fly over to Singapore to learn this insightful heuristic of modeling. Save your time and money; and most importantly, gain a competitive edge over your peers in mastering this visualization strategy in mathematical problem solving in just a few hours!

Learn what creative and innovative local math teachers and tutors are teaching their above-average students. Update and upgrade yourself with commonly used problem-solving strategies, such as the Stack Model Method to solve nonroutine and challenging questions.

Disappointingly, because of math editors' ignorance of the Stack Model Method, and poor marketing in promoting it, Singapore-published textbooks and workbooks, not to say, supplementary math titles, have yet to be incorporated with this intuitive problem-solving strategy. Moreover, you are unlikely to learn the Stack Model Method from Singapore math workshops and conferences any time soon, or even from problem-solving courses offered to trainee or in-service teachers.

Stack up your visualization skills by empowering yourself with the Singapore's stack model method! Like the Singapore's bar model method, the stack model method will help you to enhance your visual literacy or visualization skills. The time to arm yourself mathematically and professionally is NOW!

As a Singapore math educator, coach, or parent, don't shortchange yourself by learning only a fraction of the Singapore math curriculum; learn other problem-solving strategies other than the Singapore model (or bar) method, which have until now not been made readily available in print and online, especially among those outside Singapore. Here's your chance to learn the stack model method, a more powerful visualization strategy than the bar model method, which could help you and your students enhance both your creative and lateral thinking skills in elementary mathematics.

Happy creative problem solving!

Kow-Cheong Yan
kcyan.mathplus@gmail.com

Contents

Preface	iii
A (*Very Short*) History of the Stack Model Method	7
Heuristic: Draw a diagram	9
No *Venn-ture*, No gain	10
Bar Model Method vs. Stack Model Method	11
Are we born-again STACKERS?	13
One way to l@@k at it	14
Visualization Problem-Solving Strategies	15
What the Stack Method IS NOT	16
A devil's definition	17
A "Before-After" grade 4 question	18
Singapore Math 2.0	19
Basic Tools for the Stack Method	20
Whole Numbers	22
Something to think about... The Stack Method (I)	42
Age Problems	43
Fractions	51
Something to think about... The Stack Method (II)	76
Ratio & Proportion	77
Something to think about... The Stack Method (III)	105
Percentage	106
Something to think about... The Stack Method (IV)	117
Gap and Differences Concept	118
Something to think about... The Stack Method (V)	132
Revision Questions	133
Answers & Solutions	160
Bibliography & References	183
About the Author	184

A (*Very Short*) History of the *Stack Model Method*

Unlike the Singapore's "Model (or Bar) Method" that formally appeared in the mid-eighties, the "Stack Model Method," based on one or two self-published books, seems to have made its public appearance in Singapore about a decade ago.

Even today, a large proportion of local school math teachers aren't familiar with this powerful visualization problem-solving strategy, as compared to their counterparts teaching in enrichment and tuition centers, who started introducing it to grades 5–6 students, as an alternative method of solution to the bar model method.

Why the Stack Model Method has failed to reach out to a wider audience earlier is two-fold: (a) editors' unfamiliarity with the method; and (b) poor marketing efforts to promote the problem-solving strategy in local schools. In other words, local publishers are clueless about publishing books on it, because their editorial staff were mostly non-math majors, with a superficial understanding of even the bar method. In fact, about five years ago, I started submitting some solutions to word problems, which lend themselves to the Stack Model Method, but editors then conveniently rejected them—they showed zero interest to wanting to know how the visualization strategy works, or just refused to be mathematically or visually challenged.

Because the Stack Model Method was initially exemplified to solve grades 5–6 questions, elementary math teachers mistakenly thought that this problem-solving strategy is only relevant to those teaching at these grades. In fact, the Stack Model Method can also be applied to solve nonroutine questions in grades 3–4, or even at lower grades, especially when it comes to solving challenging or olympiad math questions in grades 1–2.

In recent years, thanks to social media, it has become easier for local bloggers, writers, and teachers to share with fellow math educators online why the Stack Model Method is no longer an optional visualization problem-solving strategy, reserved only for the better or above-average math students. There have been several requests to come up with a quick-and-dirty book on the Stack Model Method, but it was hard to publish it inexpensively, especially when local "math editors" (and designers and typesetters) aren't prepared to have an intuitive feel of it.

After many excuses to put off writing a book on the Stack Model Method, and following more inquiries in recent years from math educators overseas, who are keen to learn more about this intuitive and creative problem-solving method, I finally decided to write two quick-and-dirty books on this visualization strategy. The first book focuses on grades 3–4 questions; the second book looks at grades 4–6 questions, which are stack-model-friendly.

In the last revised editions of my local grades 3–6 assessment books, I was encouraged to witness that my math editors this time round are more open and willing to incorporating, when space permits, a few alternative solutions that lend themselves to the stack model method. I am confident that the stack model method will be as popular as the bar model method in coming years, as both local and foreign math educators are convinced why the stack model method is a powerfully more intuitive and creative problem-solving strategy than the bar model method, especially when it comes to solving word problems that are bar-model-unfriendly.

Heuristic: Draw a diagram

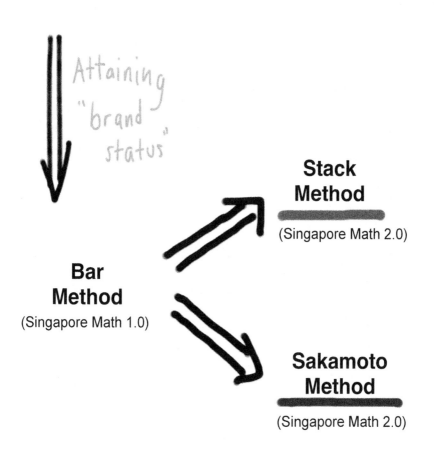

No Venn-ture, No gain

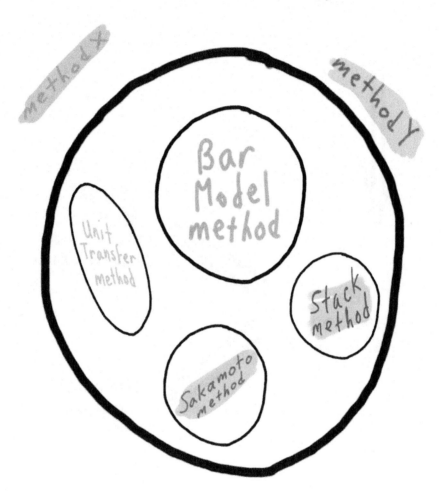

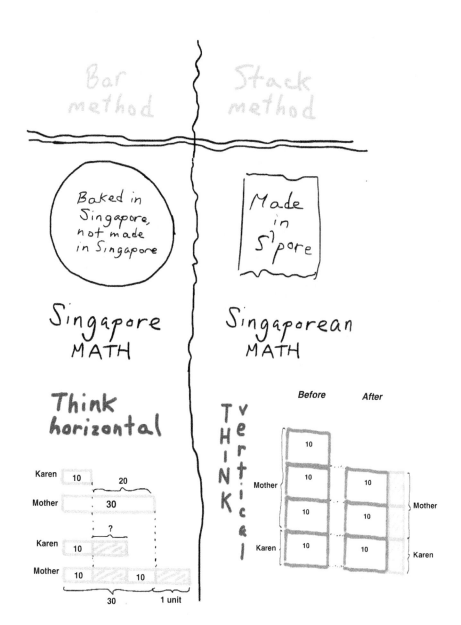

Bar method	Stack method
Hit SG ~30 years	< 10 years old
Out of age	5 years late
Look-see proofs for ADULTs	Look-see proofs for KIDS
Reading a textbook (rows)	**Reading a magazine (columns)**

Are we not born-again STACKERS?

Simplistically speaking, the stack model method may be likened to adding a string of numbers vertically, as compared to the bar model method, which involves adding the same group of numbers horizontally.

$$12.34 + 567.8 + 9.01 = ?$$

$$\begin{array}{r} 12.34 \\ 567.8 \\ +9.01 \\ \hline ? \end{array}$$

Which method is easier?

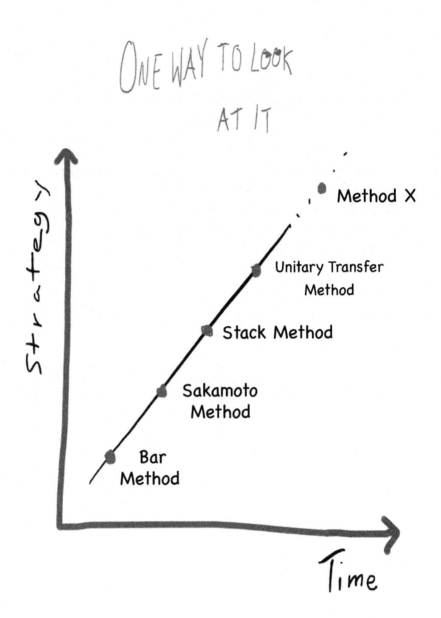

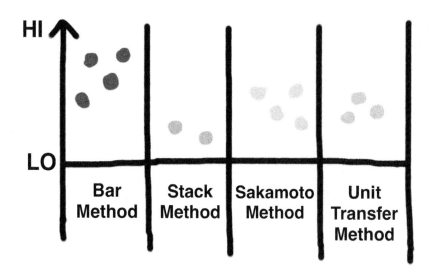

What the Stack Method IS NOT

A STACK model *is not* an inverted BAR Model

A devil's definition

Stack Model Method [math]: A problem-solving strategy that has been held hostage by pseudo-math editors in the last decade, with Singapore math authors and bloggers pushing for its release.

URL: www.stackmath.com

TWITTER: #stackmath

Futuring Robert Frost

Two model methods lay before me,
and I took the one less modeled.
And that has made all the difference.

Let's look at a "before-after" grade 4 question, taken from a Singapore math assessment book.

Rick had 3 times as much money as Fiona. After Rick gave $285 to Fiona, he had twice as much money as she did. How much money did Rick have at first?

First, solve the above word problem, using the bar model method, then compare your solution with the stack-model solution below. *Can you figure out how the method works?*

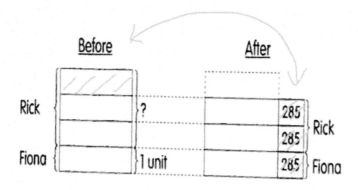

From the model,

1 unit ⟶ 3 × 285

3 units ⟶ 3 × 3 × 285 = 2565

Rick had **$2565** at first.

Note: The total amount of money remains unchanged.

Singapore Math 2.0
From Bar Method to Stack Method

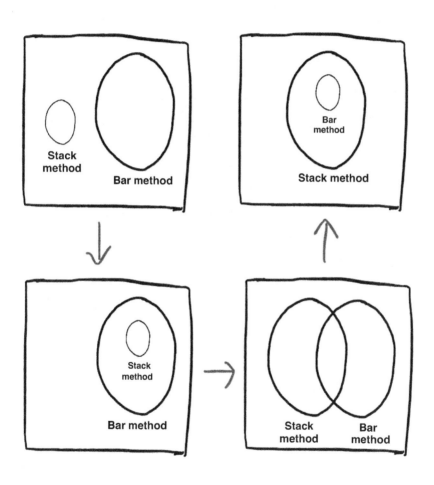

Basic Tools for
The Stack Method

Tools Needed:

Lines, Rectangles, or Squares (and a flexible brain)

Stack models may be drawn horizontally or vertically—not diagonally or circularly yet!

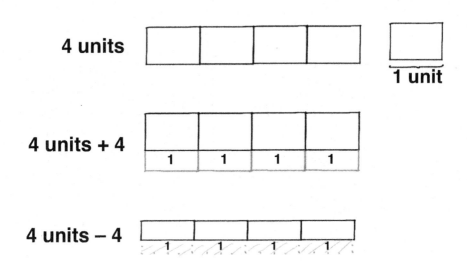

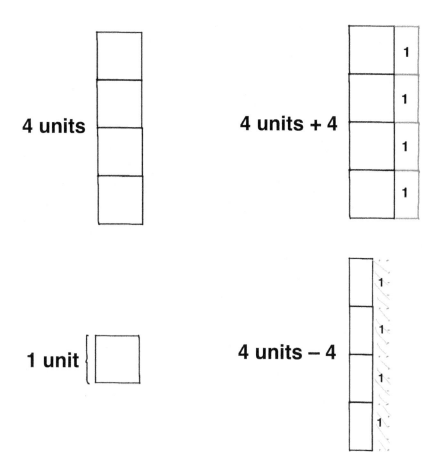

At first glance, it looks as if the stack model method is just the bar model method being rotated ninety degrees, but nothing is further from the truth. Although the bar model method and the stack model method share a fair bit in common, however, the thought process is different in each case.

Whole Numbers

Whole Numbers—Worked Examples 1–9

Here, the easier versions of the so-called pre-*before-after* word problems are exemplified, using the stack model method. These nonroutine problems would be quite a challenge without the use of algebra; however, the use of bar- or stack-modeling makes the solving of these questions look like a routine procedure.

Compare how the stack-model solutions in a number of cases offer a more intuitive (and creative) approach than the bar-model solutions in solving these word problems.

A number of these *Whole Numbers* questions make use of the same principles, or problem-solving skills, needed to solve age-related questions, as discussed in the next section.

Worked Example 1

Jim had $190 and Dawn has $60. After each of them received an equal amount of money from their father, Jim had twice as much money as Dawn. How much did their father give each of them?

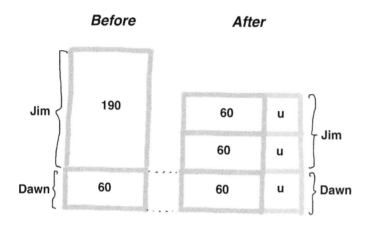

From the model,

190 = 60 + 60 + u
190 = 120 + u
u = 190 − 120 = 70

Their father gave $70 to each of them.

Note: The difference between Jim's money and Dawn's money before and after receiving the same amount of money from their father doesn't change. This is similar to the general principle that the age difference between any two persons at any point in time remains unchanged.

Thought Process

Jim had $190 and Dawn has $60. After each of them received an equal amount of money from their father, Jim had twice as much money as Dawn. How much did their father give each of them?

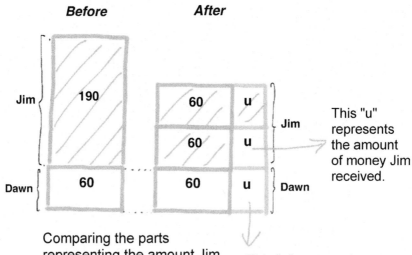

Comparing the parts representing the amount Jim had before receiving money from his father, in the before and after cases, we have:

190 = 60 + 60 + u
190 = 120 + u
u = 190 − 120 = 70

Their father gave $70 to each of them.

Worked Example 2

Mr. Yan has 12 Twitter followers and 66 Facebook friends. After gaining the same number of Twitter followers and Facebook friends, there are 3 times as many Facebook friends as Twitter followers. How many Facebook friends did he gain?

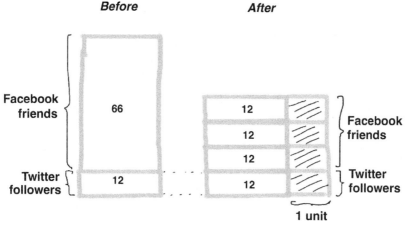

From the model,
12 + 12 + 12 + 2 units → 66
2 units → 66 − 12 − 12 − 12 = 30
1 unit → 30 ÷ 2 = 15

Mr. Yan gained 15 Facebook friends.

Check:

Before *After*
Facebook: 66   Facebook: 66 + 15 = 81
Twitter: 12   Twitter: 12 + 15 = 27

27 × 3 = 81 ✓

Thought Process

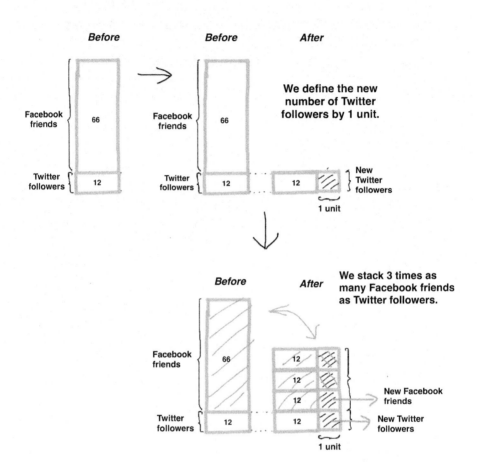

Practice

Joanne had 28 Facebook fans and Gina had 157 Facebook fans. After each of them gained the same number of fans the following week, Gina had 4 times as many fans as Joanne. How many Facebook fans did each person gain?

Try to solve this question, using both bar- and stack-modeling. Then compare your solutions. *Which one is more elegant?*

Answer: 15 fans

Worked Example 3

Paul had 40 Twitter followers and James had 128 Twitter followers. After the same number of people unfollowed each of them, James had 5 times as many Twitter followers left as Paul. How many Twitter followers unfollowed both of them?

Method 1

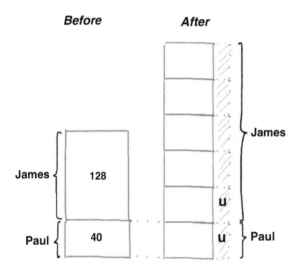

From the model,

$4u \rightarrow 5 \times 40 - 128$
$4u \rightarrow 200 - 128 = 72$
$u \rightarrow 72 \div 4 = 18$

18 Twitter followers unfollowed each of them.

$2 \times 18 = 36$
36 Twitter followers unfollowed both of them.

Method 2

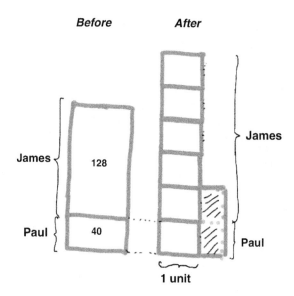

The difference in the number of Twitter followers between them will remain the same.

From the model,

4 units → 128 − 40
4 units → 88
1 unit → 88 ÷ 4 = 22
40 − 1 unit → 40 − 22 = 18

18 Twitter followers unfollowed each of them.

2 × 18 = 36
36 Twitter followers unfollowed them.

Method 3

In general, the stack method offers a more intuitive or faster method than the bar method; however, in some cases, the model method may prove to be a more effective method.

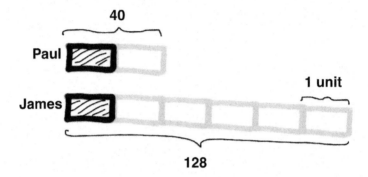

From the model,

4 units → 128 − 40 = 88
1 unit → 88 ÷ 4 = 22
40 − 1 unit → 40 − 22 = 18

18 Twitter followers unfollowed each of them.

2 × 18 = 36
36 Twitter followers unfollowed both of them.

Check:

40 − 18 = 22; 128 − 18 = 110
22 × 5 = 110

Lack-and-Shortage Questions

Worked Example 4

Rita and her friends wanted to give a farewell gift to their math teacher. If each student paid $8, there would be $3 left. However, if each of them paid $7, they would be short of $4.
(a) How many students were there?
(b) How much was the gift worth?

Let one unit (☐) represent the number of students.

Since the price of the gift does not change, we can equate the prices for different amounts of contribution as follows:

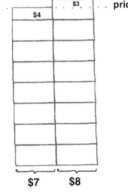

price of gift

Comparing prices:

7 units + $4 = 8 units − $3

From the model,

1 unit → 4 + 3 = 7
So there are 7 students.

7 units + 4 → 7 × 7 + 4 = 49 + 4 = 53
or
8 units − 3 → 8 × 7 − 3 = 56 − 3 = 53

The gift cost $53.

Practice

Mrs. Smith takes her students to an island. If each boat takes only 4 students, 15 students will be left behind. If each boat takes only 6 students, 5 students will be left behind.
(a) How many boats are there?
(b) How many students are there?

Answer: 5 boats; 35 students.

Worked Example 5

At a tree-planting campaign, each student is expected to plant an equal number of trees. If each student plants 5 trees, there will be 11 trees left. If each student plants 7 trees, there will be 3 trees left.
(a) How many students take part in the green campaign?
(b) How many trees are being planted?

Method 1

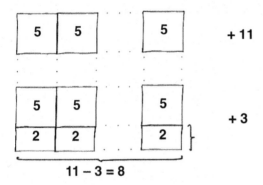

From the model,

7 − 5 = 2
There is a difference of 2 trees.

8 ÷ 2 = 4
There are 4 students.

(b) 4 × 5 + 11 = 31 or 4 × 7 + 3 = 31
31 trees are being planted.

Worked Example 5

At a tree-planting campaign, each student is expected to plant an equal number of trees. If each student plants 5 trees, there will be 11 trees left. If each student plants 7 trees, there will be 3 trees left.

(a) How many students take part in the green campaign?
(b) How many trees are being planted?

Method 2

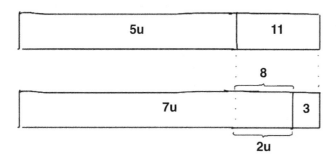

From the model, if u represents the number of students, then

2u → 11 − 3 = 8
u → 4

There are 4 students.

(b) 5 × 4 + 11 = 31 or 7 × 4 + 3 = 31
    31 trees are being planted.

Practice

A number of candies are to be shared among a group of students. If each student received 7 candies, there would be 5 candies left. If each student received 9 candies, there would be short of 29 candies.
(a) How many students are there?
(b) How many candies are there?

Answer: (a) 17 (b) 124.

Worked Example 6

James had $2900 and Rose had $1000 at first. After James had bought a laptop that cost twice as much as the laptop Rose bought, James has now 5 times as much money left as Rose. How much did Rose's laptop cost?

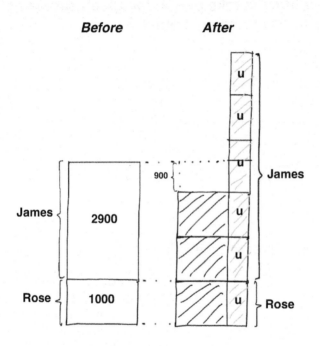

From the model,

3u → 2900 − 2 × 1000 = 900
u → 900 ÷ 3 = 300
1000 − u → 1000 − 300 = 700

Rose's laptop cost $700.

Thought Process

James had $2900 and Rose had $1000 at first. After James had bought a laptop that cost twice as much as the laptop Rose bought, James has now 5 times as much money left as Rose. How much did Rose's laptop cost?

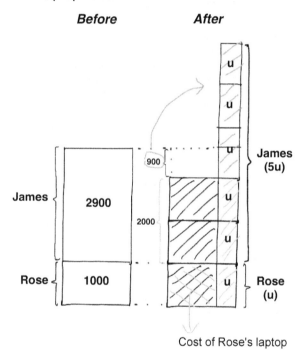

From the model,

3u → 900
u → 900 ÷ 3 = 300
1000 − u → 1000 − 300 = 700

Rose's laptop cost $700.

Worked Example 7

Carl had $400 and Esther had $120 at first. The book Carl bought cost $30 more than twice the book Esther bought. Carl has now four times as much money left as Esther has left. How much did Carl pay for his book?

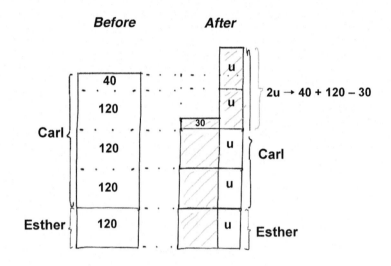

From the model,

2u → 40 + 120 – 30 = 130
u → 130 ÷ 2 = 65
4u → 4 × 65 = 260
400 – 4u → 400 – 260 = 140

Carl paid $140 for his book.

Thought Process

Carl had $400 and Esther had $120 at first. The book Carl bought cost $30 more than twice the book Esther bought. Carl has now four times as much money left as Esther has left. How much did Carl pay for his book?

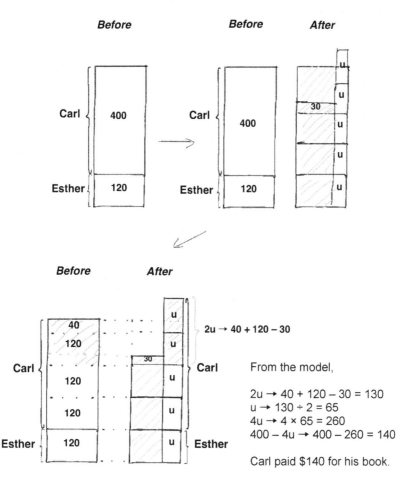

$2u \rightarrow 40 + 120 - 30$

From the model,

$2u \rightarrow 40 + 120 - 30 = 130$
$u \rightarrow 130 \div 2 = 65$
$4u \rightarrow 4 \times 65 = 260$
$400 - 4u \rightarrow 400 - 260 = 140$

Carl paid $140 for his book.

Worked Example 8

Mark had $210 and Rick had $30 at first. After receiving the same amount of money from their father, Mark has now four times as much money as Rick. How much money did each person receive?

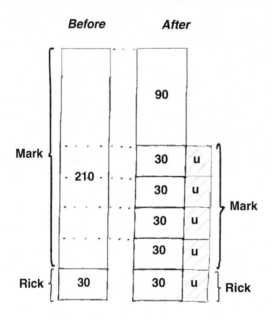

From the model,

3u → 90
u → 90 ÷ 3 = 30

Each person received $30.

Thought Process

Mark had $210 and Rick had $30 at first. After receiving the same amount of money from their father, Mark has now four times as much money as Rick. How much money did each person receive?

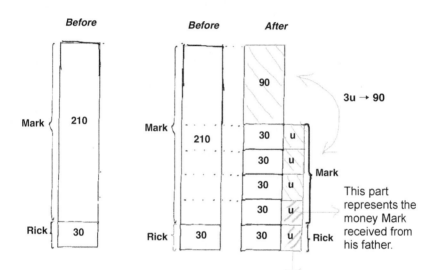

This part represents the money Rick received from his father.

From the model,

3u → 90
u → 90 ÷ 3 = 30

Each person received $30.

Worked Example 9

On Saturday, telco shops A and B processed a total of 228 broadband subscriptions. On Sunday, after shop A processed another 40 new subscriptions and shop B another 12 new subscriptions, shop A had now four times as many subscriptions as shop B. How many broadband subscriptions did each shop process at first?

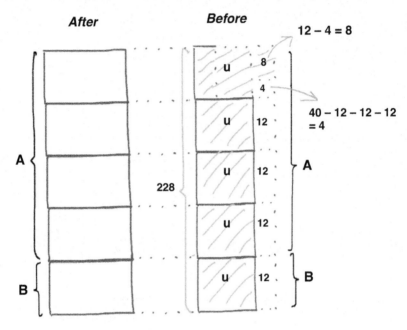

From the model,

5u → 228 − 8 = 220
u → 220 ÷ 5 = 440 ÷ 10 = 44

Shop B processed 44 subscriptions at first.

228 − 44 = 184
Shop A processed 184 subscriptions at first.

Thought Process

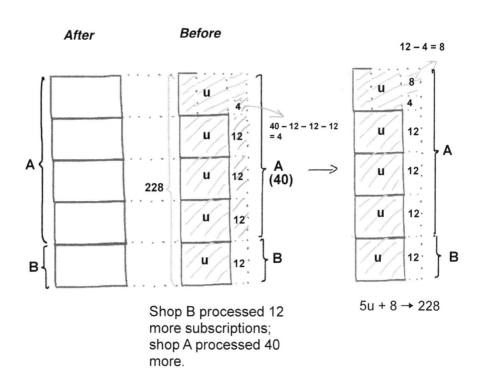

Shop B processed 12 more subscriptions; shop A processed 40 more.

Something to think about ...
The Stack Method

The next best thing to flying over to Singapore and attending a few sessions on the Stack Model Method (and being a few thousands bucks poorer), is to apply what you can learn from this book to enhance your visual literacy.

I thought I knew more than what is often taught at Singapore math conferences and symposiums. How wrong I was when the Stack Model Method reveals so much I needed to know about the power of model diagrams in solving word problems.

If bar modeling has so far mesmerized you visually, stack modeling would probably blow your mind's eye! It's "Singapore Math 2.0."

The "Stack Model Method" is like a cookbook for creative model drawing—the creative or intuitive use of stack models in solving word problems.

Stack modeling is a disruptive methodology in visualization: it is a far more effective problem-solving strategy than bar modeling, because it is creatively and intuitively richer!

If you want to be a "black sheep" among the mathematical brethren, the Stack Method could make you stand out from the crowd of mediocre problem solvers.

Age Problems

Age Problems — Worked Examples 10–13

In grades 7–8, age-related problems are popular questions that are often set to test the algebraic skills of students. However, thanks to bar- and stack-modeling, these same questions could be given to grades 5–6 students who haven't formally been introduced to solving algebraic equations.

Interestingly, a number of these age word problems can prove quite resistant to the bar model method, but are favorable to the stack model method.

Worked Example 10

Mrs. Gan is 5 times as old as her godson. In 18 years' time, she will be only twice as old as him. How old is Mrs. Gan now?

Method 1

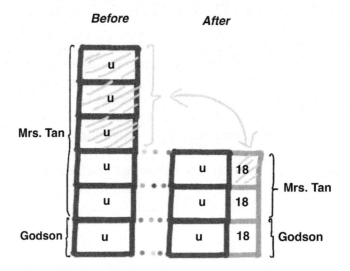

From the model,

3u → 18
u → 18 ÷ 3 = 6
5u → 5 × 6 = 30

Mrs. Gan is 30 years old now.

Check

Mrs. Gan	Godson	
30	6	6 × 5 = 30
+ 18	+ 18	
48	24	24 × 2 = 48

Thought Process

Mrs. Gan is 5 times as old as her godson. In 18 years' time, she will be only twice as old as him. How old is Mrs. Gan now?

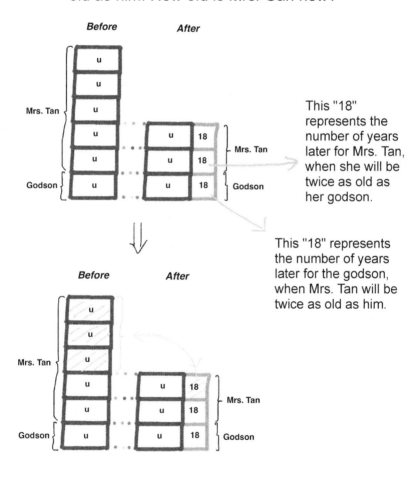

This "18" represents the number of years later for Mrs. Tan, when she will be twice as old as her godson.

This "18" represents the number of years later for the godson, when Mrs. Tan will be twice as old as him.

Mrs. Gan is 5 times as old as her godson. In 18 years' time, she will be only twice as old as him. How old is Mrs. Gan now?

Method 2

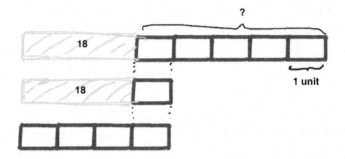

From the model,

3 units → 18
1 unit → 18 ÷ 3 = 6
5 units → 5 × 6 = 30

Mrs. Gan is 30 years old now.

Practice

Mr. Yan is 3 times as old as his daughter. Six years ago, he was 6 times as old as her. How old will Mr. Yan be 3 years from now? Answer: 33 years old.

Worked Example 11

Six years ago, Dorine's mother was 6 times as old as her. Her mother is now 3 times as old as Dorine. How old is Dorine's mother now?

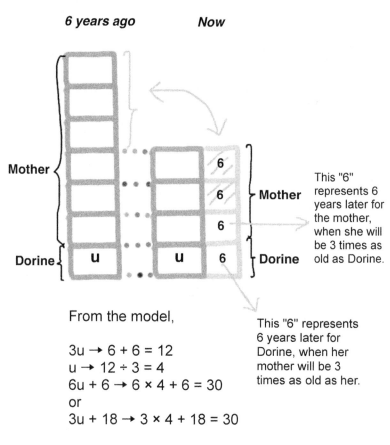

From the model,

$3u \rightarrow 6 + 6 = 12$
$u \rightarrow 12 \div 3 = 4$
$6u + 6 \rightarrow 6 \times 4 + 6 = 30$
or
$3u + 18 \rightarrow 3 \times 4 + 18 = 30$

Dorine's mother is now 30 years old.

Worked Example 12

The total age of Mr. Oliver and his son is 72 years. Mr. Oliver is 6 years older than twice his son's age. How many years ago was Mr. Oliver five times as old as his son?

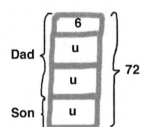

From the model,
3u → 72 – 6
u → 24 – 2 = 22
72 – 22 = 50 or 2u + 6 = 44 + 6 = 50

The son is 22 years old and the father is 50 years old.

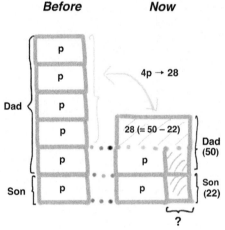

From the model,
4p → 28
p → 7

The son was 7 years old and the father was 35 years old, when Mr. Oliver was 5 times as old as his son.

22 – p → 22 – 7 = 15
15 years ago, Mr. Oliver was 5 times as old as his son.

Thought Process

The total age of Mr. Oliver and his son is 72 years.
Mr. Oliver is 6 years older than twice his son's age.
How many years ago was Mr. Oliver five times as old as his son?

From the model,

3u → 72 – 6 = 66
u → 66 ÷ 3 = 22
72 – 22 = 50 or 2u + 6 = 44 + 6 = 50

The son is 22 years old and the father is 50 years old.

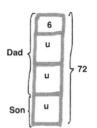

Before   **Now**

From the model,

4p → 28
p → 7

The son was 7 years old and the father was 35 years old, when Mr. Oliver was 5 times as old as his son.

22 – p → 22 – 7 = 15
15 years ago, Mr. Oliver was 5 times as old as his son.

Note: From the models, there is no need to use the fact that the age difference between father and son doesn't change at any point in time to solve the problem. However, making use of this fact does help to solve the problem faster, but that sounds more like "using a formula" to solve the question.

This part represents the number of years ago for Mr. Oliver when he was 5 times as old as his son.

This part represents the number of years ago for the son when his father was 5 times as old as him.

Worked Example 13

Henry's age 4 years ago was 3 times as old as his son's age in 6 years' time. The sum of Henry's age in 5 years' time and his son's age 3 years ago was 44. How old is Henry now?

The sum of Henry's age in 5 years' time and his son's age 3 years ago was 44.

The sum of Henry's age and his son's age now is:
44 − 5 + 3 = 42.

Now

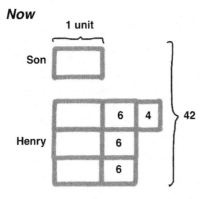

If the son's age now represents 1 unit, in 6 years' time, his age will be "1 unit + 6."

Three times "1 unit + 6" represents "3 units + 6 + 6 + 6," which was the age of Henry 4 years ago.

So, Henry's age now must be "3 units + 6 + 6 + 6 + 4."

From the model,

4 units → 42 − 6 − 6 − 6 − 4 = 20
1 unit → 20 ÷ 4 = 5
42 − 1 unit → 42 − 5 = 37

Henry is 37 years old now.

Practice

Jerry's age 5 years ago was 3 times as old as his son's age in 4 years' time. The sum of Jerry's age in 3 years' time and his son's age 7 years ago was 45. How old is Jerry now? Answer: 41 years old.

Fractions

Fractions — Worked Examples 14–23

In grades 5–6, nonroutine questions on *Fractions* could easily be tweaked to be posed as questions on *Ratio and Proportion*, or *Percentage*, as most use the same mathematical principles to solving them.

As no formal knowledge of algebraic techniques is required at this level, at least for local students in Singapore, and because a number of these challenging word problems have proved favorable to the stack model method, this has allowed pre-algebra students to lay hold of these nonroutine, higher-order thinking questions at a younger age.

Worked Example 14

Four years ago, Terry's age was 2/5 his father's age. In 6 years' time, Terry will be half as old as his father. How old are the son and father now?

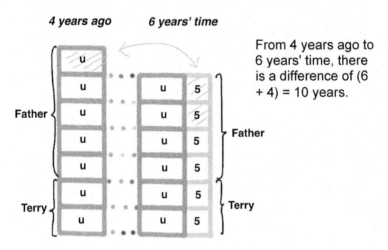

4 years ago   *6 years' time*

From 4 years ago to 6 years' time, there is a difference of (6 + 4) = 10 years.

From the model,

$u \to 5 + 5 = 10$
$4u + 20 - 6 \to 40 + 14 = 54$

or $5u + 4 \to 50 + 4 = 54$
The father is 54 years old.

$2u + 4 \to 20 + 4 = 24$
The son is 24 years old.

Try solving the question using the bar model method. Clearly, the stack model method "outwits" the bar model method.

Thought Process

Four years ago, Terry's age was 2/5 his father's age. In 6 years' time, Terry will be half as old as his father. How old are the son and father now?

From 4 years ago to 6 years' time, there is a difference of (6 + 4) = 10 years.

Terry's age → 2/5 father's age
Before: Terry's age → 2 units
        Father's age → 5 units

Terry's age → 1/2 of father's age
After: Terry's age → 2 units + 10 = 2 units + 5 + 5
       Father's age → 4 units + 20

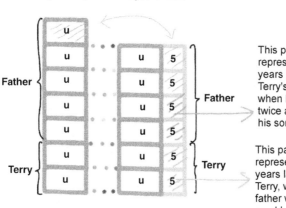

From the model,

u → 5 + 5 = 10
4u + 20 − 6 → 4 × 10 + 14 = 40 + 14 = 54

or 5u + 4 → 5 × 10 + 4 = 50 + 4 = 54
The father is 54 years old.

2u + 4 → 20 + 4 = 24
The son is 24 years old.

Worked Example 15

Robin is 3/5 as old as her mother. In 12 years, Robin will be 2/3 as old as her mother.
a) How old is Robin now?
b) How old was her mother when Robin was born?

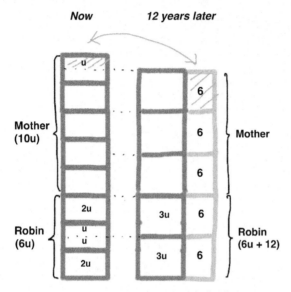

From the model,

u → 6
6u → 6 × 6 = 36

Robin is 36 years old now.

The age difference between them is 10u − 6u = 4u = 4 × 6 = 24 years, which remains unchanged at any point in time.

This means that when Robin was born, her mother was 24 years old.

Thought Process

Robin is 3/5 as old as her mother. In 12 years, Robin will be 2/3 as old as her mother.
a) How old is Robin now?
b) How old was her mother when Robin was born?

Before: Robin is 3/5 as old as her mother.
Robin's age → 3 units; mother's age → 5 units, or
Robin's age → 6 units; mother's age → 10 units.

After: Robin will be 2/3 as old as her mother. Robin's age: 6 units + 12; mother's age → 9 units + 18

Note: 9 units + 18 → 10 units + 12, or 1 unit → 6.

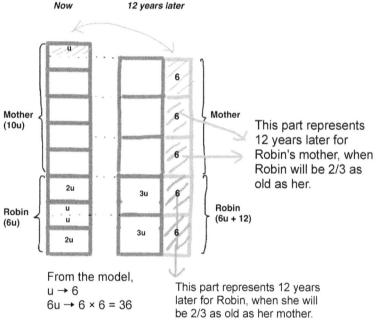

From the model,
u → 6
6u → 6 × 6 = 36

This part represents 12 years later for Robin, when she will be 2/3 as old as her mother.

Robin is 36 years old now.

The age difference between them is 10u − 6u = 4u = 4 × 6 = 24 years, which remains unchanged at any point in time.

This means that when Robin was born, her mother was 24 years old.

Worked Example 16

The age of Roy is 3/7 of the age of his father. 14 years later, his age will be 4/7 of the age of his father. How old is Roy now?

Method 1

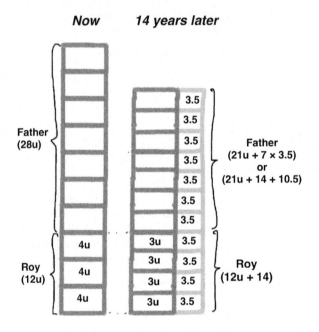

From the model,

$28u + 14 \rightarrow 21u + 14 + 3 \times 3.5$
$28u - 21u \rightarrow 3 \times 3.5$
$7u \rightarrow 10.5$
$u \rightarrow 1.5$
$12u \rightarrow 12 \times 1.5 = 18$

Roy is 18 years old now.

Method 2

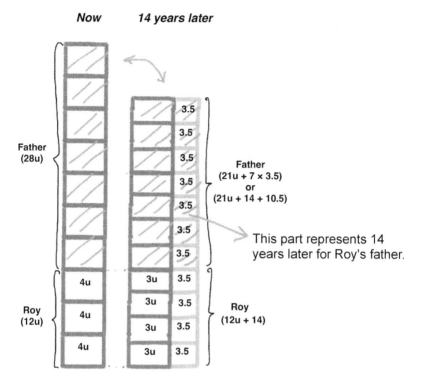

From the model,

28u → 21u + 3 × 3.5
28u − 21u → 10.5
7u → 10.5
u → 1.5
12u → 12 × 1.5 = 18

Roy is 18 years old now.

A Bar- Or Stack-Unfriendly Question?

Worked Example 17

The age of Rose is 4/7 the age of her mother. In 10 years' time, she will be 7/11 as old as her mother. How old is Rose now?

Some age problems do not lend themselves easily to the bar model method, or sometimes even to the stack model method. The above question is one such example, where the bar model may be too complicated to be drawn, or it is just too cumbersome to represent the given information in a model drawing. When this happens, other methods, such as the Sakamoto method or the unitary (or units) method, are preferred.

Using the units method

Before
Rose: 4 units × 4 = 16 units
Mother: 7 units × 4 = 28 units

After
Rose: 7 units × 3 = 21 units
Mother: 11 units × 3 = 33 units

Increase in units = 21 − 16 = 5 or 33 − 28 = 5
5 units → 10 years
1 unit → 10 ÷ 5 = 2 years
16 units → 16 × 2 = 32 years

Rose is 32 years old now.

Note: We do some guess and check to both the "before" and "after" cases, until the difference in units between Rose and her mother in either case is the same, which is 12 units.

Method 2

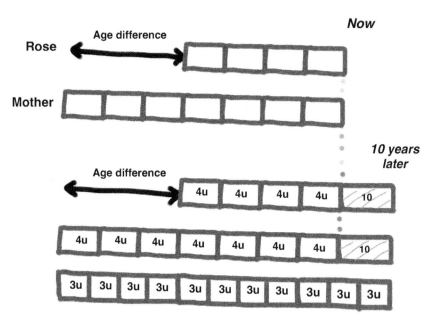

From the model,

4 × 4u + 10 → 7 × 3u
16u + 10 → 21u = 16u + 5u
5u → 10
u → 2
16u → 16 × 2 = 32

or

4u + 10 → 3 × 3u = 9u
4u + 10 → 4u + 5u
5u → 10
u → 2

Rose is 32 years old now.

Note: It takes a while to make sense of the bar models in order to elicit any meaningful numerical relationship.

Method 3 (Optional)

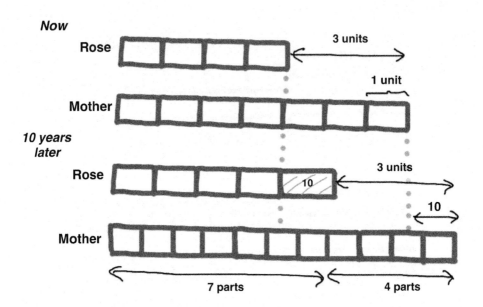

From the model,

4 parts = 3 units
1 part = 3/4 unit

Also, 7 parts = 4 units + 10
7 × 3/4 unit = 4 units + 10
21/4 units − 4 units = 10
(21/4 − 16/4) units = 10
5/4 units = 10
1 unit = 10 × 4/5 = 8
4 units = 4 × 8 = 32

Rose is 32 years old now.

Note: Method 3 looks like a pseudo-bar model drawing, which often appears in a number of assessment books. Formal algebraic thinking lurks behind the method of solution.

Worked Example 18

Mrs. Yan had 400 grapefruit and mangoes. She sold 3/4 of the grapefruit and 2/3 of the mangoes. Then she had 120 grapefruit and mangoes left. What fraction of the fruits were mangoes at first?

Method 1

400 − 360 = 40

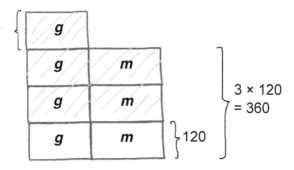

From the model,

$g \rightarrow$ 400 − 3 × 120 = 400 − 360 = 40

Given: $g + m \rightarrow$ 120
$m \rightarrow$ 120 − g = 120 − 40 = 80
$3m \rightarrow$ 3 × 80 = 240

Fraction of fruits that were mangoes at first = 240/400 = 3/5

Worked Example 18

Mrs. Yan had 400 grapefruit and mangoes. She sold 3/4 of the grapefruit and 2/3 of the mangoes. Then she had 120 grapefruit and mangoes left. What fraction of the fruits were mangoes at first?

Method 2

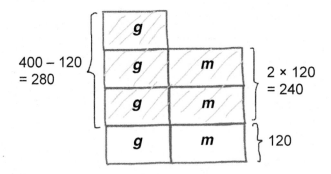

From the model,

$g \rightarrow 280 - 240 = 40$
$m \rightarrow 120 - g = 120 - 40 = 80$
$3m \rightarrow 3 \times 80 = 240$

Fraction of fruits that were mangoes at first
= 240/400 = 3/5

TH!NK: *Can you draw a third stack model that allows you to form a new relationship between the grapefruit and the mangoes?*

Questions 19 and 20—Same, Yet Different

Worked Example 19

A total of 52 people logged in to a free live webinar. After 3/4 of the men and 3/5 of the women logged off, there was an equal number of men and women remaining. How many women logged in to the webinar at first?

Since the remaining number of men is equal to the number of women, 1/4 of the men must be equal to 2/5 of the women.

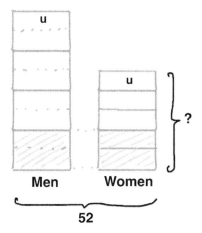

From the model,

total number of units = 8 + 5 = 13
13 units → 52
1 unit → 52 ÷ 13 = 4
5 units → 5 × 4 = 20

20 women logged in to the webinar at first.

Worked Example 19

A total of 52 people logged in to a free live webinar. After 3/4 of the men and 3/5 of the women logged off, there was an equal number of men and women remaining. How many women logged in to the webinar at first?

Thought Process

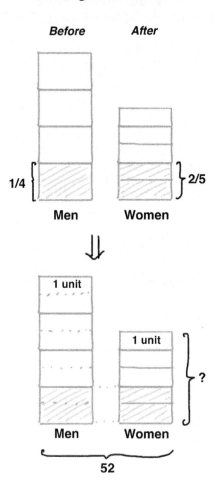

Worked Example 20

A total of 64 people logged in to a free live webinar. After 3/4 of the men and 3/5 of the women logged off, the total number of men and women who remained was 19. How many women logged in to the webinar at first?

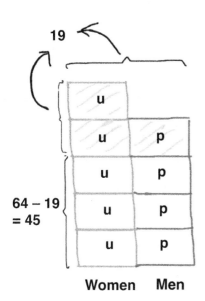

From the model,

$3u + 3p \rightarrow 64 - 19 = 45$
$u + p \rightarrow 45 \div 3 = 15$

$2u + p \rightarrow 19$
$u \rightarrow 19 - 15 = 4$
$5u \rightarrow 5 \times 4 = 20$

The number of women who logged in at first was 20.

Worked Example 21

Mr. Peak had 48 more Facebook friends than Facebook fans. After 5/6 of his Facebook friends and 3/4 of his Facebook fans unfollowed him, he still had 33 online friends and fans left.

a) How many Facebook fans did he have at first?
b) What fraction of those who unfollowed him were Facebook friends?

Method 1

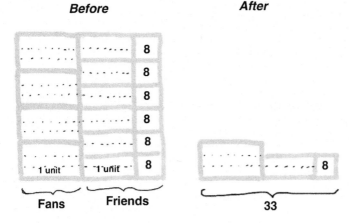

From the model,
5 units → 33 − 8 = 25
1 unit → 25 ÷ 5 = 5
12 units → 12 × 5 = 60

Mr. Peak had 60 Facebook fans at first.

Number of fans who unfollowed him = 3/4 × 60 = 45

Number of Facebook friends = 60 + 48 = 108
Number of friends who unfollowed him = 5/6 × 108 = 90

Fraction required = 90/(90 + 45) = 90/135 = 2/3

Therefore, 2/3 who unfollowed Mr. Peak were Facebook friends.

Thought Process

Mr. Peak had 48 more Facebook friends than Facebook fans. After 5/6 of his Facebook friends and 3/4 of his Facebook fans unfollowed him, he still had 33 online friends and fans left.
a) How many Facebook fans did he have at first?
b) What fraction of those who unfollowed him were Facebook friends?

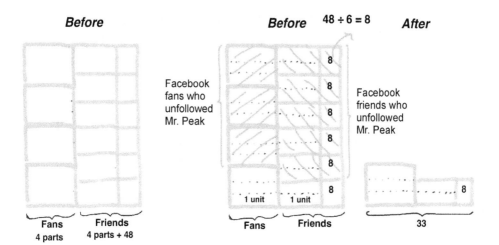

We look for a common unit that divides the 4 parts representing the fans and the 6 new parts representing the friends. This can be found by taking the least common multiple of 4 and 6, which is 12.

From the model,

5 units → 33 − 8 = 25
1 unit → 25 ÷ 5 = 5
12 units → 12 × 5 = 60

Mr. Peak had 60 Facebook fans at first.

Alternatively, we may use "u" to represent one unit.

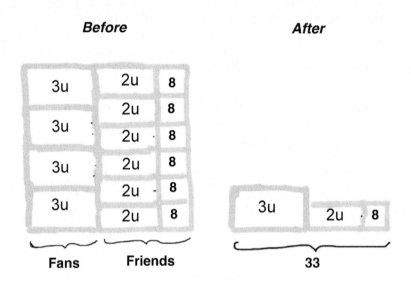

From the model,

5u → 33 − 8 = 25
u → 25 ÷ 5 = 5
12u → 12 × 5 = 60

Mr. Peak had 60 Facebook fans at first.

Worked Example 21

Mr. Peak had 48 more Facebook friends than Facebook fans. After 5/6 of his Facebook friends and 3/4 of his Facebook fans unfollowed him, he still had 33 online friends and fans left.
a) How many Facebook fans did he have at first?
b) What fraction of those who unfollowed him were Facebook friends?

Method 2

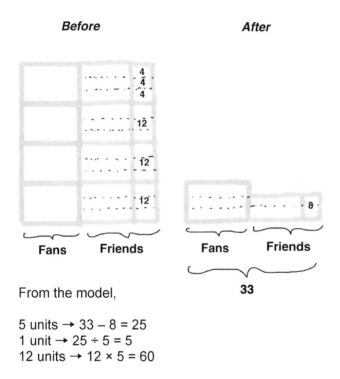

From the model,

5 units → 33 − 8 = 25
1 unit → 25 ÷ 5 = 5
12 units → 12 × 5 = 60

Mr. Peak had 60 Facebook fans at first.

Thought Process

Mr. Peak had 48 more Facebook friends than Facebook fans. After 5/6 of his Facebook friends and 3/4 of his Facebook fans unfollowed him, he still had 33 online friends and fans left.
a) How many Facebook fans did he have at first?
b) What fraction of those who unfollowed him were Facebook friends?

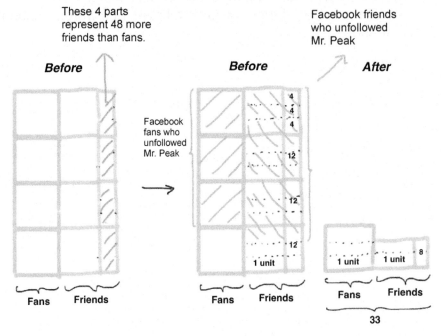

From the model,

5 units → 33 − 8 = 25
1 unit → 25 ÷ 5 = 5
12 units → 12 × 5 = 60

Mr. Peak had 60 Facebook fans at first.

The number of fans is expressed in 4 parts, and the number of friends is given in 6 parts.

We subdivide both models into a common unit, such that it is easier to make comparison between friends and fans. To do this, we take the least common multiple of 4 and 6, which is 12. So, we divide the 4 parts for both friends and fans into 12 units.

Worked Example 22

Fanny had five times as many cards as Roy. Fanny gave 1/4 of her cards to Roy. Then, Roy gave 1/6 of his cards to Fanny in return. In the end, Fanny had 90 more cards than Roy. How many cards did Fanny have at first?

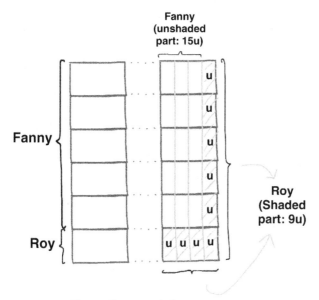

From the model,

Roy: 1/6 × 9u = 1½u; 9u − 1½u = 7½u
Fanny: 15u + 1½u = 16½u

16½u − 7½u → 90
9u → 90
u → 10
20u → 20 × 10 = 200

Fanny had 200 cards at first.

Thought Process

Fanny had five times as many cards as Roy. Fanny gave 1/4 of her cards to Roy. Then, Roy gave 1/6 of his cards to Fanny in return. In the end, Fanny had 90 more cards than Roy. How many cards did Fanny have at first?

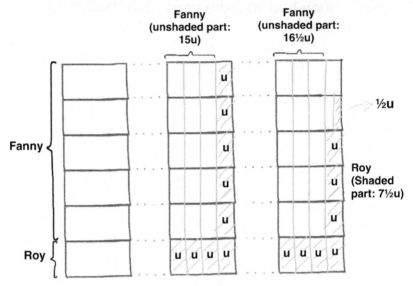

Roy → 4u; Fanny → 5 × 4u
Fanny gave 1/4 of her cards to Roy → 5u
Roy had now (4u + 5u) = 9u, and Fanny was left with (20u − 5u) = 15u.

Roy gave 1/6 of his cards to Fanny, which is equal to 1/6 × 9u = 3u/2. So, Roy was left with 9u − 3u/2 = 15u/2.

Fanny had now (15u + 3u/2) = 33u/2.

33u/2 − 15u/2 → 90
9u → 90
u → 10

Or, from the model drawing (unshaded part), we could cross out 7½ units to produce 9 remaining units.

Worked Example 23

Samuel and Jill had 576 stamps in all. Samuel gave 1/7 of his stamps to Jill. Then, Jill gave 1/4 of the total number of stamps she now had to Samuel. In the end, both of them had the same number of stamps.
(a) How many stamps did Samuel have at first?
(b) What was the difference in the number of stamps they each had at first?

Method 1

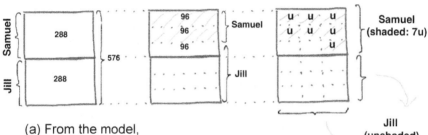

(a) From the model,

576 ÷ 6 = 96
3u → 96
u → 96 ÷ 3 = 32
7u → 7 × 32 = 224

Samuel had 224 stamps at first.

(b) 576 − 224 = 352
Jill had 352 stamps at first.

352 − 224 = 128
The difference in the number of stamps is 128.

Or, cross out 7 units from the unshaded units, leaving only 4 units: 4 × 32 = 128

The difference is 128.

Thought Process

Samuel and Jill had 576 stamps in all. Samuel gave 1/7 of his stamps to Jill. Then, Jill gave 1/4 of the total number of stamps she now had to Samuel. In the end, both of them had the same number of stamps.
(a) How many stamps did Samuel have at first?
(b) What was the difference in the number of stamps they each had at first?

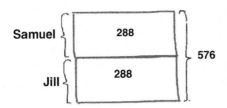
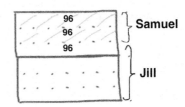

Since the total number of stamps remains unchanged, and Samuel and Jill each had the same number of stamps, this means that in the end, each person would be left with 576 ÷ 2 = 288.

After Jill gave 1/4 of her stamps to Samuel, they each had the same number of stamps. This means that before the transfer, Jill had 4 parts and Samuel had 2 parts. Also, 3 parts must be equal to 288, or 1 part equals 96.

After Samuel gave 1/7 of his share to Jill, he had 2 parts left, while Jill must have 4 parts minus 1/7 of Samuel's part before.

In other words, at first, Samuel must have 7 units, with 2 remaining parts equal to 6 units, after giving 1 unit to Jill.

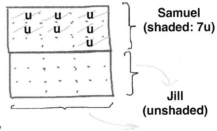

Also, 3 units → 96
    1 unit → 96 ÷ 3 = 32
    7 units → 7 × 32 = 224

Samuel had 224 stamps at first.

Worked Example 23

Samuel and Jill had 576 stamps in all. Samuel gave 1/7 of his stamps to Jill. Then, Jill gave 1/4 of the total number of stamps she now had to Samuel. In the end, both of them had the same number of stamps.
(a) How many stamps did Samuel have at first?
(b) What was the difference in the number of stamps they each had at first?

Method 2

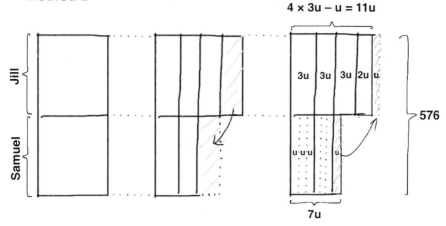

From the model,

At first, Jill → 11u; Samuel → 7u

Now, 11u + 7u → 576
          18u → 576
             u → 576 ÷ 18 = 32

Samuel: 7u → 7 × 32 = 224
Samuel had 224 stamps at first.

Difference: 11u − 7u = 4u
            4u → 4 × 32 = 128
The difference in the number of stamps is 128.

Something to think about ...
The Stack Method

Like the bar model method, the stack model method enables word problems that were traditionally set at higher grades to be set at lower grades. Its intuitive and visual appeal makes the stack model method a problem-solving strategy of choice among visual learners!

A side-effect of the Stack Method is that a working knowledge of it could help raise your "Visualization Quotient" (VQ) by at least a dozen points!

Unlike bar modeling that toggles sideways, stack modeling can take place in a left-right and an up-down manner. The thought (or design) process often needs creativity and lateral thinking. Often times, there is more than "one right way" to construct the stack drawing.

Like the model method, the first idea that comes to mind to constructing the stack model is very unlikely to be the best. That is why it is so important to go on thinking and to produce other possible stack models.

Not all stack models are created equal—some are more elegant than others.

An elegant stack model may not necessarily be easier to understand, especially for the novice—it may be visually beautiful but conceptually harder to "see." A trade-off is often required: understanding or elegance; comprehensiveness or simplicity.

Ratio & Proportion

Ratio and Proportion—Worked Examples 24–35

As pointed out in the previous section, questions on *Ratio* and *Proportion* are disguised versions of questions on *Fractions*, although they may conceptually be slightly harder, as they involve one or two more steps.

Again, nonroutine questions on ratio and proportion lend themselves quite easily to the stack model method. What is interesting is that as the questions get challenging or computationally tedious, they tend to be more favorable to the stack model method than to the bar model method. Of course, this may not be obvious to the novice problem solver, who has started to make sense of both bar- and stack-modeling.

However, in some rare cases when the word problem doesn't lend itself to either bar modeling or stack modeling, then other problem-solving strategies, such as the *Unitary* and *Sakamoto* methods, could be called upon to address the situation.

Worked Example 24

Paul had twice as much money as David. After Paul and David donated money to a charity in the ratio 3 : 1, Paul had $64 left and David had $48. How much money did they donate altogether?

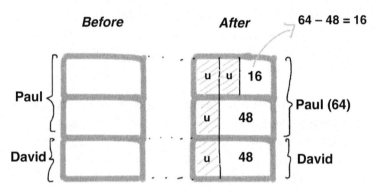

u represents the amount David donated to charity.

From the model,

u → 48 − 16 = 32
4u → 4 × 32 = 128

They donated a total of $128.

Worked Example 25

Ryan had 3 times as much money as Naomi. After Ryan and Naomi spent money buying e-books in the ratio 2 : 1, Ryan had $78 left and Naomi had $14 left. How much money did Ryan spend?

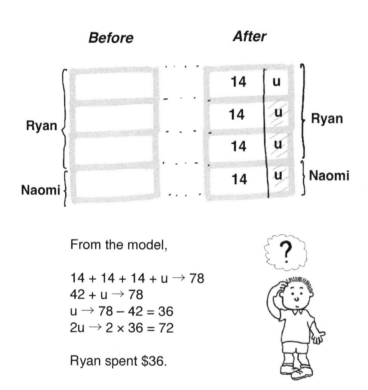

From the model,

$14 + 14 + 14 + u \rightarrow 78$
$42 + u \rightarrow 78$
$u \rightarrow 78 - 42 = 36$
$2u \rightarrow 2 \times 36 = 72$

Ryan spent $36.

Thought Process

Ryan had 3 times as much money as Naomi. After Ryan and Naomi spent money buying e-books in the ratio 2 : 1, Ryan had $78 left and Naomi had $14 left. How much money did Ryan spend?

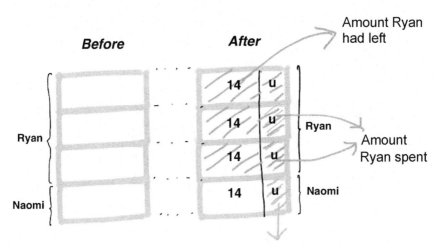

From the model,

14 + 14 + 14 + u → 78
42 + u → 78
u → 78 − 42 = 36
2u → 2 × 36 = 72

Ryan spent $36.

Worked Example 26

Lisa weighs one-and-a-half times as heavy as Pearl. After Lisa had lost 10 kg and Pearl had gained 12 kg, the ratio of Lisa's weight to Pearl's weight became 5 : 4. What was Lisa's weight at first?

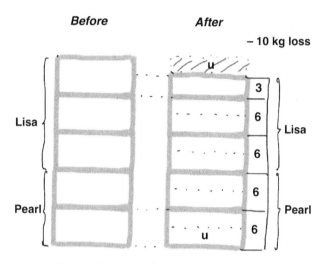

From the model,

u → 10 + 3 + 6 + 6 = 25
6u → 6 × 25 = 150

Lisa weighed 150 kg at first.

Check:
Lisa: 150 − 10 = 140
Pearl: 4u → 4 × 25 = 100; 100 + 12 = 112
140 : 112 = 5 × 28 : 4 × 28 = 5 : 4

Thought Process

Lisa weighs one and a half times as heavy as Pearl. After Lisa had lost 10 kg and Pearl had gained 12 kg, the ratio of Lisa's weight to Pearl's weight became 5 : 4. What was Lisa's weight at first?

Lisa weighs 1½ times as heavy as Pearl.
1½ = 3/2
Pearl → 3 parts; Lisa → 2 parts

The ratio of Lisa's weight to Pearl's weight was 5 : 4.
Pearl → 4 units; Lisa → 5 units

1 part → 6 kg, or 2u → 6 kg
½ part → 3 kg, or u → 3 kg

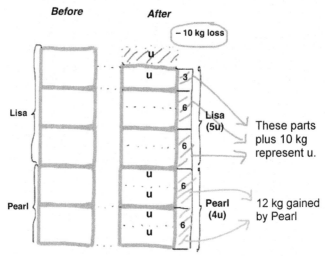

From the model,
u → 10 + 3 + 6 + 6 = 25
6u → 6 × 25 = 150

Lisa weighed 150 kg at first.

Check:
Lisa: 150 − 10 = 140
Pearl: 4u → 4 × 25 = 100; 100 + 12 = 112
140 : 112 = 5 × 28 : 4 × 28 = 5 : 4

Worked Example 27

The ratio of the number of #SGmath hashtags to the number of #SG50 hashtags Ben used was 4 : 7. Had he used 48 less #SGmath hashtags, the ratio would have been 2 : 5 instead. How many #SGmath hashtags were used?

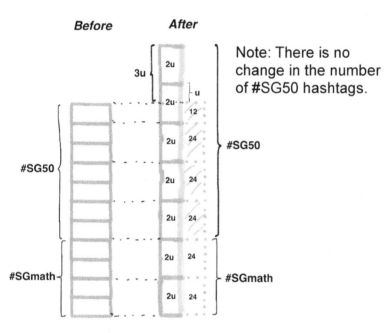

From the model,

$3u \to 24 + 24 + 24 + 12$
$u \to 8 + 8 + 8 + 4 = 28$
$4u + 48 \to 4 \times 28 + 48 = 112 + 48 = 160$

Therefore, 160 #SGmath hashtags were used.

Thought Process

The ratio of the number of #SGmath hashtags to the number of #SG50 hashtags Ben used was 4 : 7. Had he used 48 less #SGmath hashtags, the ratio would have been 2 : 5 instead. How many #SGmath hashtags were used?

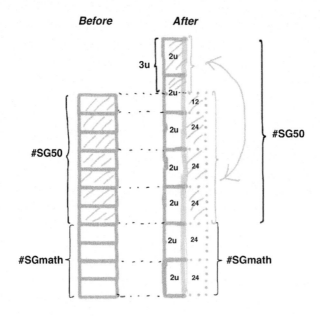

2 : 5 = 4 : 10

    Before   After
4 parts : 7 parts → 4 units : 10 units

Since there is no change in the number of #SG50 hashtags, the 7 parts representing them before must be equal to the 10 units representing them after.

From the model, the 3 units must be equal to "24 + 24 + 24 + 12."

Note: 2u → 24
       u → 12

Worked Example 28

The number of hours Ian spent texting to the number of hours tweeting is 5 : 3. After spending 6 more hours tweeting, the ratio becomes 2 : 3. How many hours did he text?

Method 1

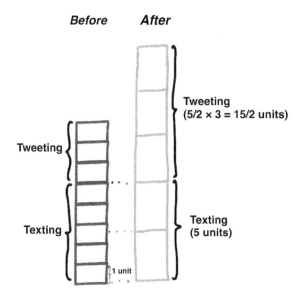

From the model,

15/2 units − 3 units → 6 hours
9/2 units → 6 hours
1 unit → 6 × 2/9 = 4/3 hours
5 units → 5 × 4/3 hours = 20/3 hours
= 6⅔ hours

Ian texted for 6⅔ hours.

Thought Process

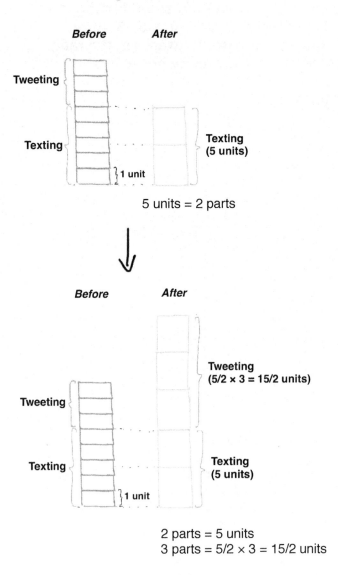

Worked Example 28

The number of hours Ian spent texting to the number of hours tweeting is 5 : 3. After spending 6 more hours tweeting, the ratio becomes 2 : 3. How many hours did he text?

Method 2

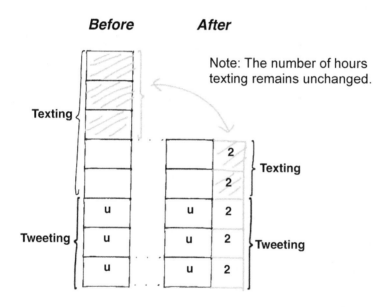

From the model,

3u → 2 + 2 = 4
u → 4/3
5u → 5 × 4/3 = 20/3 = 6⅔

Ian texted for 6⅔ hours.

Thought Process

The number of hours Ian spent texting to the number of hours tweeting is 5 : 3. After spending 6 more hours tweeting, the ratio becomes 2 : 3. How many hours did he text?

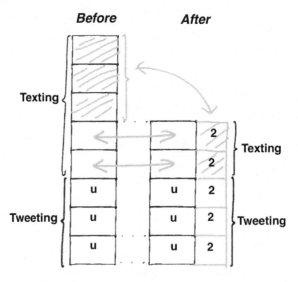

Since the number of hours texting remains unchanged,

5 units → 2 units + 2 + 2
3 units → 2 + 2

This result can directly be obtained from the model, by comparing the shaded parts, or eliminating common parts.

3 units → 2 + 2 = 4
1 unit → 4/3

Different Contexts, Same Principle

Worked Example 29

The ratio of the number of Joan's coins to the number of Roy's coins was 3 : 4. After Roy gave 28 coins to his cousin, and Joan received 36 coins from her father, the ratio of the number of Joan's coins to the number of Roy's coins became 6 : 7. How many coins Joan had in the beginning?

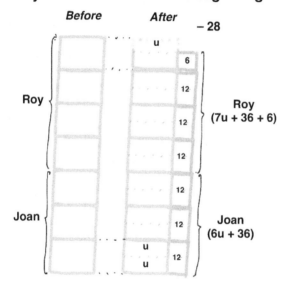

From the model,

u → 28 + 6 + 12 + 12 + 12 = 70
6u → 6 × 70 = 420

Joan has 420 coins at first.

	Joan	Roy
Before	6u → 420	8u → 560
After	420 + 36 = 456	560 − 28 = 532

456 : 532 = 6 × 76 : 7 × 76 = 6 : 7

Thought Process

The ratio of the number of Joan's coins to the number of Roy's coins was 3 : 4. After Roy gave 28 coins to his cousin, and Joan received 36 coins from her father, the ratio of the number of Joan's coins to the number of Roy's coins became 6 : 7. How many coins Joan had in the beginning?

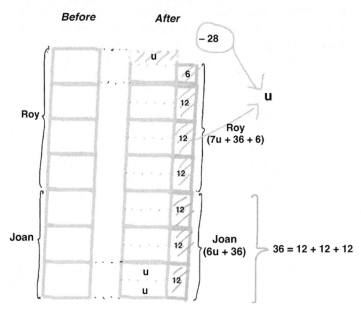

Before
3 parts : 4 parts = 6 units : 8 units
Joan → 6 units
Roy → 8 units

After
Joan → 6 units + 36
Roy → 7 units + 42

Comparing Roy's parts:
8 units – 28 → 7 units + 42
8 units – 7 units → 42 + 28
1 unit → 70
6 units → 6 × 70 = 420
Joan has 420 coins at first.

From the model,

u → 28 + 6 + 12 + 12 + 12 = 70
6u → 6 × 70 = 420

Joan has 420 coins at first.

Worked Example 30

The ratio of the number of Sally's postcards to the number of Dave's postcards was 2 : 3. After Sally had sold 24 postcards and Dave had bought 18 postcards, the ratio of the number of Sally's postcards to the number of Dave's postcards became 4 : 9. How many postcards did Dave have at first?

Method 1

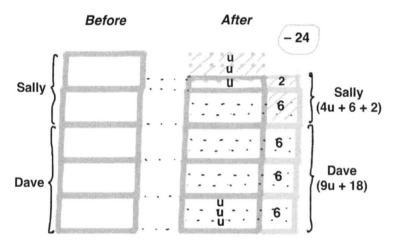

From the model,

2u → 24 + 2 + 6 = 32
u → 32 ÷ 2 = 16
9u → 9 × 16 = 144

Dave had 144 postcards at first.

Check

	Dave	Sally
Before	9u → 144	6u → 96
After	144 + 18 = 162	92 − 24 = 72

72 : 162 = 4 × 18 : 9 × 18 = 4 : 9

Thought Process

The ratio of the number of Sally's postcards to the number of Dave's postcards was 2 : 3. After Sally had sold 24 postcards and Dave had bought 18 postcards, the ratio of the number of Sally's postcards to the number of Dave's postcards became 4 : 9. How many postcards did Dave have at first?

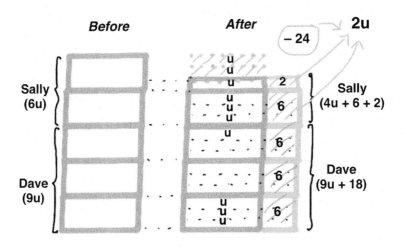

Before
Sally : Dave = 2 : 3
              = 6 units : 9 units

After
Sally : Dave = 4 parts : 9 parts

For Dave, 9 [shaded] parts represent 18 postcards.

9 parts → 18
3 parts → 6
1 part → 2

From the model,

2u → 24 + 2 + 6 = 32
u → 32 ÷ 2 = 16
9u → 9 × 16 = 144

Dave had 144 postcards at first.

Worked Example 30

The ratio of the number of Sally's postcards to the number of Dave's postcards was 2 : 3. After Sally had sold 24 postcards and Dave had bought 18 postcards, the ratio of the number of Sally's postcards to the number of Dave's postcards became 4 : 9. How many postcards did Dave have at first?

Method 2

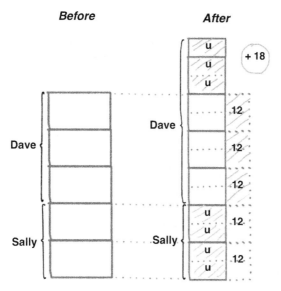

From the model,

$3u \rightarrow 12 + 12 + 12 + 18$
$u \rightarrow 4 + 4 + 4 + 6 = 18$
$6u + 3 \times 12 \rightarrow 6 \times 18 + 36 = 108 + 36 = 144$
or $9u - 18 \rightarrow 9 \times 18 - 18 = 8 \times 18 = 144$

Dave had 144 postcards at first.

Thought Process

The ratio of the number of Sally's postcards to the number of Dave's postcards was 2 : 3. After Sally had sold 24 postcards and Dave had bought 18 postcards, the ratio of the number of Sally's postcards to the number of Dave's postcards became 4 : 9. How many postcards did Dave have at first?

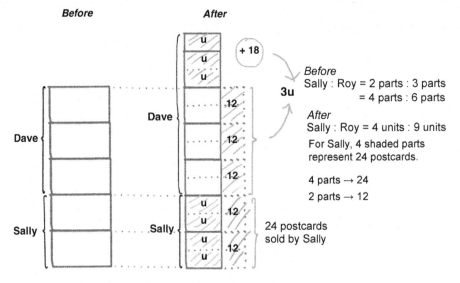

From the model,

3u → 12 + 12 + 12 + 18
u → 4 + 4 + 4 + 6 = 18
6u + 3 × 12 → 6 × 18 + 36 = 108 + 36 = 144
or 9u − 18 → 9 × 18 − 18 = 8 × 18 = 144

Dave had 144 postcards at first.

Worked Example 31

An egg seller had eggs in baskets P, Q, and R in the ratio 4 : 2 : 3. She transferred eggs from baskets P and Q to basket R in the ratio of 4 : 1 respectively. As a result, the number of remaining eggs in basket P and in basket Q was 48 and 32, respectively. How many eggs were there in basket R in the end?

Method 1

P : Q : R
4 : 2 : 3
2 : 1

48 − 32 = 16   32 − 16

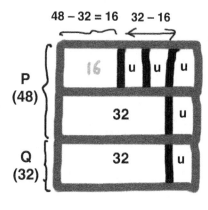

u → number of eggs taken from basket Q to basket R

From the model,

2u → 32 − 16 = 16
u → 16 ÷ 2 = 8

Q → 32 + 8 = 40
Basket Q has 40 eggs at first.

Since Q : R = 2 : 3,
R = $^3/_2$ Q = $^3/_2$ × 40 = 60
Basket R has 60 eggs at first.

Now, R + 5u → 60 + 5 × 8
= 60 + 40
= 100

In the end, basket R has 100 eggs.

Worked Example 31

An egg seller had eggs in baskets P, Q, and R in the ratio 4 : 2 : 3. She transferred eggs from baskets P and Q to basket R in the ratio of 4 : 1 respectively. As a result, the number of remaining eggs in basket P and in basket Q was 48 and 32 respectively. How many eggs were there in basket R in the end?

Method 2

P : Q : R
4 : 2 : 3
2 : 1

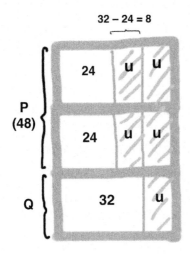

u → number of eggs taken from basket Q to basket R
4u → number of eggs taken from basket P to basket R

From the model,

$48 \div 2 = 24$

u → 32 − 24 = 8
5u → 5 × 8 = 40

Q → 32 + 8 = 40
Basket Q has 40 eggs at first.

Since Q : R = 2 : 3,
R = $\frac{3}{2}$ Q = $\frac{3}{2}$ × 40 = 60
Basket R has 60 eggs at first.

New R = R + 5u
= 60 + 40
= 100

Basket R has 100 eggs in the end.

Worked Example 32

The ratio of Henry's money to Jane's money was 3 : 2. After Henry gave away $7 and Jane received $12, the ratio of Henry's money to Jane's money became 5 : 4. How much money did Jane have at first?

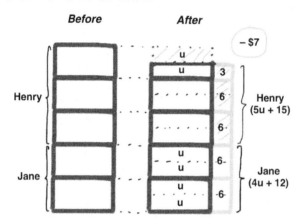

From the model,

u → $7 + $3 + $6 + $6 = $22
4u → 4 × $22 = $88

Jane had $88 at first.

Check

Before
Henry: 6u → 6 × 22 = 132
Jane: 88
Henry : Lily = 132 : 88 = 3 : 2

After
Henry: 132 − 7 = 125 = 5 × 25
Jane: 88 + 12 = 100 = 4 × 25
Henry : Jane = 5 : 4

Worked Example 33

Celine gives 24 coins to Jeffrey, and the ratio of Jeffrey's coins to Celine's coins is now 2 : 3. If Celine has given Jeffrey 36 coins, the ratio of Jeffrey's coins to Celine's coins will be 6 : 7. How many coins does Celine have in the beginning?

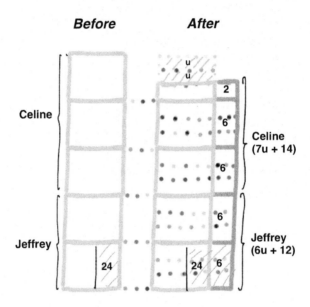

From the model,

2u → 2 + 4 × 6 = 26
u → 26 ÷ 2 = 13
9u → 9 × 13 = 117
9u + 24 → 117 + 24 = 141

Celine has 141 coins at first.

Thought Process

Celine gives 24 coins to Jeffrey, and the ratio of Jeffrey's coins to Celine's coins is now 2 : 3. If Celine has given Jeffrey 36 coins, the ratio of Jeffrey's coins to Celine's coins will be 6 : 7. How many coins does Celine have in the beginning?

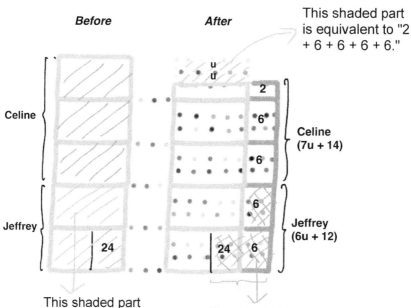

This shaded part belongs to Jeffrey.

Before
Jeffrey : Celine = 2 : 3
Jeffery → 2 parts − 24
Celine → 3 parts + 24

After
Jeffrey : Celine = 6 : 7
Jeffrey → 6 units + 12
Celine → 7 units + 14

This shaded part represents the 36 coins (= 24 + 6 + 6) given to Jeffrey.

From the model,

$2u → 2 + 4 × 6 = 26$
$u → 26 ÷ 2 = 13$
$9u → 9 × 13 = 117$
$9u + 24 → 117 + 24 = 141$

Celine has 141 coins at first.

Worked Example 34

Paul and John shared some stamps in a ratio of 5 : 7. After they each donated 26 stamps to a museum, the ratio became 1 : 4. How many stamps did John have at first?

Method 1

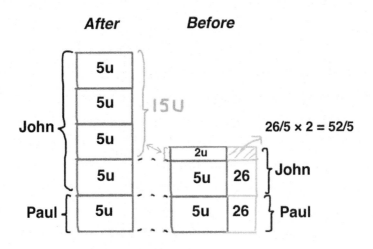

From the model,

15u − 2u → 52/5
13u → 52/5
u → 52/5 ÷ 13 = 4/5
20u → 20 × 4/5 = 16
20u + 26 → 16 + 26 = 42

John had 42 stamps at first.

Worked Example 34

Paul and John shared some stamps in a ratio of 5 : 7. After they each donated 26 stamps to a museum, the ratio became 1 : 4. How many stamps did John have at first?

Method 2

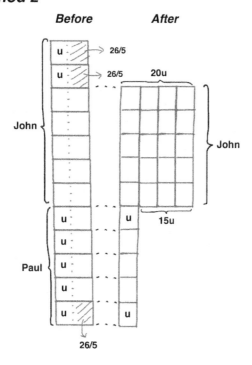

From the model,

15u − 2u → 26/5 + 26/5
13u → 52/5
u → 52/5 ÷ 13 = 4/5
20u → 20 × 4/5 = 16
20u + 26 → 16 + 26 = 42

John had 42 stamps at first.

Thought Process

Paul and John shared some stamps in a ratio of 5 : 7. After they each donated 26 stamps to a museum, the ratio became 1 : 4. How many stamps did John have at first?

Method 2

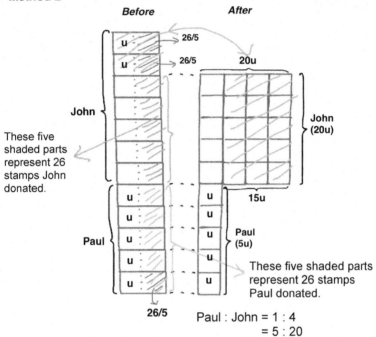

Paul : John = 1 : 4
= 5 : 20

From the model,

15u − 2u → 26/5 + 26/5
13u → 52/5
u → 52/5 ÷ 13 = 4/5
20u → 20 × 4/5 = 16
20u + 26 → 16 + 26 = 42

John had 42 stamps at first.

or

From the model,

7u + 26/5 + 26/5 → 20u
20u − 7u → 26/5 + 26/5
13u → 52/5
u → 52/5 ÷ 13 = 4/5
20u → 20 × 4/5 = 16
20u + 26 → 16 + 26 = 42

John had 42 stamps at first.

Worked Example 35

The ratio of the number of girls to the number of boys in a classroom was 6 : 5. After 16 girls left the room, the ratio became 2 : 3. How many girls were there at first?

Method 1

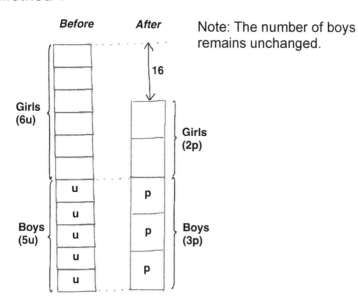

From the model,

3p → 5u
2p → 5u/3 × 2 = 10u/3

6u − 10u/3 → 16
18u/3 − 10u/3 → 16
8u/3 → 16
u → 16 × 3/8 = 6
6u → 6 × 6 = 36

There were 36 girls at first.

Worked Example 35

The ratio of the number of girls to the number of boys in a classroom was 6 : 5. After 16 girls left the room, the ratio became 2 : 3. How many girls were there at first?

Method 2 [Working backwards]

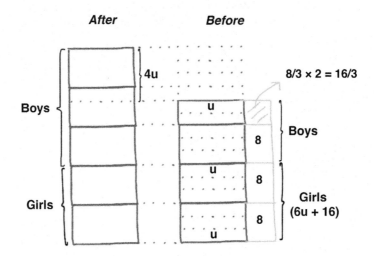

From the model,

4u → 16/3 + 8 = 40/3
u → 40/3 ÷ 4 = 10/3

6u + 16 → 6 × 10/3 + 16 = 20 + 16 = 36
There were 36 girls at first.

Note: The number of boys remains unchanged.

The dotted lines only serve as a guide to form any meaningful relationship in the before-after situations.

Something to think about ...
The Stack Method

Unlike algebra which may be performed mindlessly, stack modeling is almost impossible to fake. You need to understand the problem situation to be able to sensibly produce the correct stack drawing.

Like bar modeling, stack modeling helps you to see and to be in control of the problem situation; once you are able to draw the model, you are half-way there.

For many veteran teachers fluent in bar modeling, there seems to be few creative things left to be done to make the models more elegant; however, this isn't so with stack modeling, which lends itself more to creativity and simplicity.

Bar models are often complex and hard to follow even for a seasoned problem solver; stack models offer simpler and more intuitive models of finding the answer. Stack modeling indirectly encourages an element of simplicity, flexibility, or creativity.

For certain types of word problems, bar modeling offers an inefficient problem-solving strategy: waste of time, attention, and mental energy. Stack modeling often offers an easier and intuitive method of solution to solving challenging word problems that are bar-model-unfriendly.

Some questions may not be complicated, but the bar models needed to solve them are complex; this often generates stress, anxiety, and frustration among many average math students—stack models could provide a remedy in those cases.

Percentage

Percentage—Worked Examples 36–41

As in the case of questions on *Ratio* and *Proportion*, grades 5–6 questions on *Percentage* are disguised versions of questions on *Fraction*. *Percentage* questions often encompass other concepts from ratio, proportion, and fraction, and are often quite challenging to even above-average math students.

Not surprisingly, most of these percentage word problems lend themselves to the stack model method —they are more stack-model- than bar-model-friendly. The challenge for the seasoned problem solver is to come up with different stack models to solve the same question, and to show that some stack models are more elegant than others.

Worked Example 36

There were 60 students at a birthday party. 25% of them are girls. When more girls join the party, the percentage of girls increases to 40%. How many more girls join the party?

Method 1

25% = 1/4; 1/4 × 60 = 15
40% = 40/100 = 2/5
2/5 of students are girls → Girls : Boys = 2 : 3

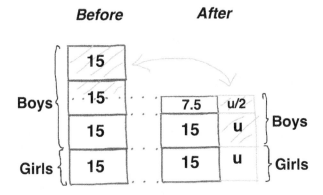

From the model,

15 + 15/2 → u + u/2
u → 15

So, 15 more girls join the party.

Worked Example 36

There were 60 students at a birthday party. 25% of them are girls. When more girls join the party, the percentage of girls increases to 40%. How many more girls join the party?

Method 2

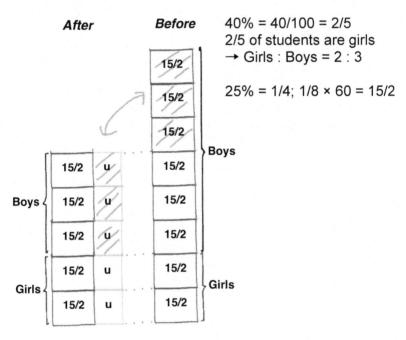

From the model,

3u → 15/2 + 15/2 + 15/2
u → 15/2
2u → 2 × 15/2 = 15

15 more girls join the party.

Worked Example 37

Kevin will earn $2400 if he sells his vintage watch set at a discount of 5% off the usual price. If he sells the watch set at a discount of 25% off the usual price, he will lose $900. What is the cost price of the vintage watch set?

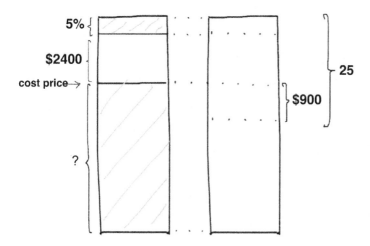

From the model,

25% − 5% → 2400 + 900 = 3300
20% → 3300
1% → 3300 ÷ 20 = 165
95% − 2400 → 95 × 165 − 2400
                 = 15,675 − 2400
                 = 13,275

The cost price of the vintage watch set is $13,275.

Worked Example 38

There were a total of 1200 tablets and smartphones in a shop, of which 60% were tablets. After selling some tablets, the percentage of tablets remaining dropped to 20%. How many tablets were sold?

Tablets: 60% = 60/100 = 3/5; smartphones: 1 − 3/5 = 2/5
Smartphones : Tablets = 2 : 3

1200 ÷ 5 = 2400 ÷ 10 = 240
Remaining tablets: 20% = 1/5

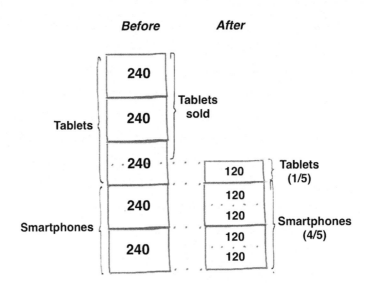

From the model,

the number of tablets sold is 240 + 240 + 120 = 600.

Worked Example 39

Pearl spent 30% of his money on some e-books and 80% of the remaining money on 14 apps. If each e-book costs 5/2 times as much as an app, how many e-books could she have bought with all her money?

30% = 30/100 = 3/10
Let Pearl's total amount represent 10 units.
Money spent on e-books = 3 units
Money remaining to spend on 14 apps = 7 units

80% = 80/100 = 4/5
Let the money remaining represent 5 parts.
Money spent on 14 apps = 4 parts
Money left in the end = 1 part

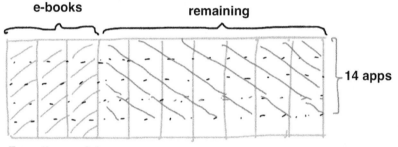

From the model,

the total amount of money represents 10 × 5 = 50 sub-units,
of which (4 × 7) = 28 sub-units represent 14 apps.

1 app represents 2 sub-units.

Each e-book costs 5/2 times as much as an app.
1 e-book represents 5/2 × 2 = 5 sub-units.

5 sub-units can buy 1 e-book.
50 sub-units can buy (50 ÷ 5) = 10 e-books.

She could have bought 10 e-books.

Thought Process

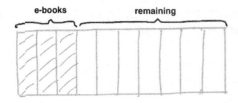

30% = 30/100 = 3/10
Pearl's total money → 10 units
Money spent on e-books → 3 units
Money remaining to spend on 14 apps → 7 units

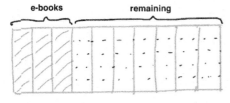

80% = 80/100 = 4/5
Money remaining → 5 parts
Money spent on 14 apps → 4 parts
Money left in the end → 1 part

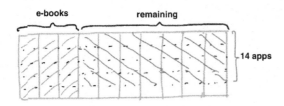

Total amount of money → 10 × 5 = 50 sub-units

14 apps → 4 × 7 = 28 sub-units
1 app → 2 sub-units

1 e-book → 5/2 apps
1 e-book → 5/2 × 2 sub-units → 5 sub-units

5 sub-units → 1 e-book
50 sub-units → 50 ÷ 5 = 10 e-books

She could have bought 10 e-books.

Worked Example 40

At a meeting, there were 40% more men than women. After 24 men and 24 women left the meeting, there were 50% more men than women. How many men were at the meeting at first?

Method 1

$40\% = 40/100 = 2/5$

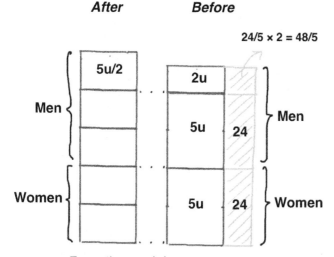

From the model,

$5u/2 - 2u \rightarrow 48/5$
$u/2 \rightarrow 48/5$
$u \rightarrow 2 \times 48/5 = 96/5$
$7.5u + 24 \rightarrow 15/2 \times 96/5 + 24 = 144 + 24 = 168$

or

$7u + 24 + 48/5 \rightarrow 7 \times 96/5 + 48/5 + 24 = 144 + 24 = 168$

There were 168 men at first.

Thought Process

At a meeting, there were 40% more men than women. After 24 men and 24 women left the meeting, there were 50% more men than women. How many men were at the meeting at first?

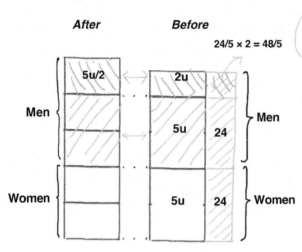

$5u \to 24$
$u \to 24/5$
$2u \to 2 \times 24/5$

$40\% = 40/100 = 2/5$

There were 2/5 more men than women.

If 5 units represent the women, then (5 + 2) = 7 units will represent the men.

At first, before 24 women left, there were (5 units + 24) women. Before 24 men left, there were [7 units + (24/5 × 7)] men.

Later, there were 50% (or 1/2) more men than women. Since the women represents 5 units (after 24 of them left), the men will represent (5 units + 1/2 × 5 units).

Comparing the models for men, we have:

$5u + 5u/2 \to 5u + 2u + 48/5$
$5u/2 \to 2u + 48/5$

Or, eliminating common parts, we have:
$5u/2 \to 2u + 48/5$, or
$5u/2 - 2u \to 48/5$
$5u/2 - 4u/2 \to 48/5$
$u/2 \to 48/5$
$u \to 2 \times 48/5 = 96/5$

Worked Example 40

At a meeting, there were 40% more men than women. After 24 men and 24 women left the meeting, there were 50% more men than women. How many men were at the meeting at first?

Method 2

40% = 40/100 = 2/5
Women : Men = 5 : 7

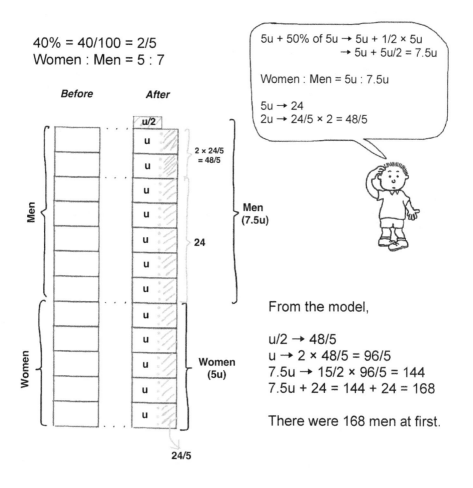

5u + 50% of 5u → 5u + 1/2 × 5u
→ 5u + 5u/2 = 7.5u

Women : Men = 5u : 7.5u

5u → 24
2u → 24/5 × 2 = 48/5

From the model,

u/2 → 48/5
u → 2 × 48/5 = 96/5
7.5u → 15/2 × 96/5 = 144
7.5u + 24 = 144 + 24 = 168

There were 168 men at first.

Worked Example 41

At a forum, 60% of the participants were men and the rest were women. When 140 more participants joined the forum, the number of men increased by 20% and the number of women increased by 40%. How many participants were there at the forum in the end?

Men: 60% = 60/100 = 3/5
Women : Men = 2 : 3

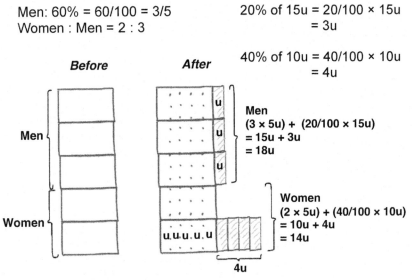

20% of 15u = 20/100 × 15u
= 3u

40% of 10u = 40/100 × 10u
= 4u

Men
(3 × 5u) + (20/100 × 15u)
= 15u + 3u
= 18u

Women
(2 × 5u) + (40/100 × 10u)
= 10u + 4u
= 14u

Note: We divide both the 3 parts for men and the 2 parts for women each into 5 smaller parts, so that it is easier to add 20% and 40%, respectively to them. This yields "1 part = 5 units."

From the model,
3u + 4u → 140
7u → 140
u → 140 ÷ 7 = 20

18u + 14u = 32u
32u → 32 × 20 = 640
There were 640 participants at the forum.

Something to think about ...
The Stack Method

It may be unfair to suggest that there is always a simpler stack model drawing. There is always the "possibility" of a simpler stack model. At times there may be no simpler way, or it may be hard to find.

Finding a simpler model is usually neither simple nor easy. However, it may be worth investing some time and effort in finding a simpler way—an elegant stack drawing isn't too far away!

Just because some less commonly known problem-solving strategies aren't formally taught in schools doesn't mean that we should shortchange our students by not introducing them to these new tools. Your math students would thank you for introducing them to the Stack Method!

Seldom do we stop to ask ourselves what types of word problems that aren't conducive, or are brain-unfriendly, to the bar model method. Are there other problem-solving visualization strategies that ease understanding for the students? Yes, the stack method is one of them! So is the Sakamoto method.

Simplicity often involves the "trading-off" of one value against another. Abstraction vs. visualization (or intuition). You gain comprehension and visualization, but lose in generalization and abstraction. Stack modeling could act as a bridge between bar modeling and algebra.

Gap and Differences Concept

Gap and Differences Concept—Worked Examples 42–48

Arguably, the so-called "gap and differences concept" questions are supposed to be the ones that often pose much difficulty even to those with a good grasp of elementary math concepts.

Since algebraic techniques is not formally taught to grades 5–6 students in Singapore, the stack model method has proved once again to be quite users-friendly to these types of questions. In fact, the stack model method has revealed itself to be a more intuitive and creative method of solution than the bar model method, when it comes to solving these types of challenging ratio questions.

Worked Example 42

If Ann gave Beth $2, Beth would have twice as much as Ann. If Beth gave Ann $2, they would have the same amount of money. How much did each person have?

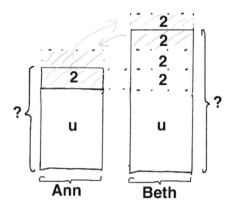

From the model,

u → 2 + 2 + 2 + 2 = 8
u + 2 → 8 + 2 = 10
u + 6 → 8 + 6 = 14

Ann had $10.
Beth had $14.

Check

Ann	Beth	Ann	Beth
10	14	10	14
− 2	+ 2	+ 2	− 2
8	16	12	12

Practice

If Ann gave Beth $2, Beth would have three times as much as Ann. If Beth gave Ann $2, they would have the same amount of money. How much did each person have?

Answer: Ann: $6; Beth: $10.

Thought Process

If Ann gave Beth $2, Beth would have twice as much as Ann. If Beth gave Ann $2, they would have the same amount of money. How much did each person have?

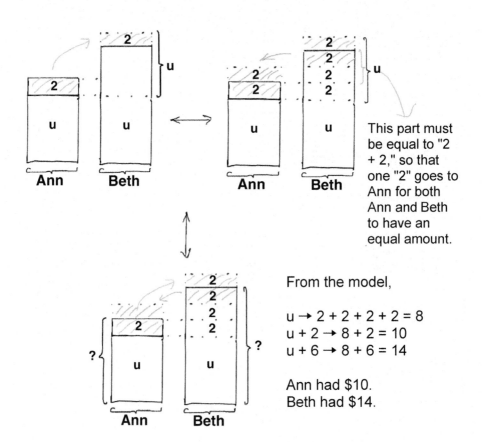

This part must be equal to "2 + 2," so that one "2" goes to Ann for both Ann and Beth to have an equal amount.

From the model,

u → 2 + 2 + 2 + 2 = 8
u + 2 → 8 + 2 = 10
u + 6 → 8 + 6 = 14

Ann had $10.
Beth had $14.

Worked Example 43

If Ann gave Beth $2, Beth would have five times as much as Ann. If Beth gave Ann $2, Ann would have the same amount of money. How much did each person have?

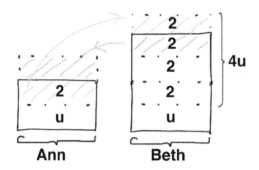

From the model,

4u → 2 + 2 + 2 + 2
u → 2
u + 2 → 2 + 2 = 4
u + 6 → 2 + 6 = 8

Ann had $4.
Beth had $8.

Check

Ann	Beth	Ann	Beth
4	8	4	8
− 2	+ 2	+ 2	− 2
2	10	6	6

Thought Process

If Ann gave Beth $2, Beth would have five times as much as Ann. If Beth gave Ann $2, Ann would have the same amount of money. How much did each person have?

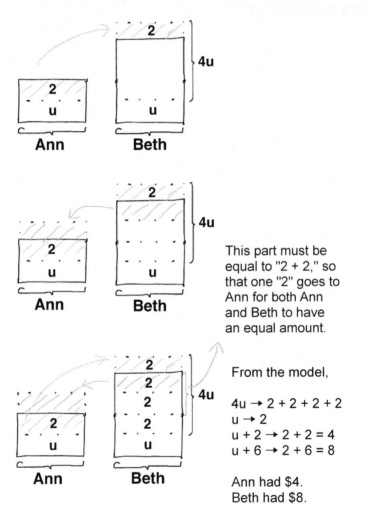

This part must be equal to "2 + 2," so that one "2" goes to Ann for both Ann and Beth to have an equal amount.

From the model,

$4u \rightarrow 2 + 2 + 2 + 2$
$u \rightarrow 2$
$u + 2 \rightarrow 2 + 2 = 4$
$u + 6 \rightarrow 2 + 6 = 8$

Ann had $4.
Beth had $8.

Practice

If Ann gave Beth $2, Beth would have three times as much as Ann. If Beth gave Ann $2, Ann would have twice as much as Beth. How much did each person have?

Solve this question, using both the Bar and Stack Model methods. Which one is easier?

Answer: Ann: $4.40; Beth: $5.20.

Worked Example 44

If Zoe gives 15 candles to Henry, she will have 3 times as many candies as him. However, if she gives 30 candles to him, she will have twice as many candies as him. How many candies does Zoe have?

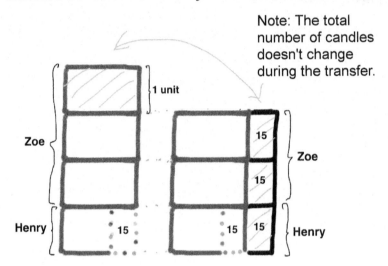

Note: The total number of candles doesn't change during the transfer.

From the model,

1 unit → 15 + 15 + 15 = 45
3 units → 3 × 45 = 135
3 units + 15 → 135 + 15 = 150

Zoe has 150 candies.

Check:

Zoe: 150
    150 − 30 = 120

Henry: 45 − 15 = 30
    30 + 30 = 60

Worked Example 45

In a game, Oliver lost 10 stamps to Jill and had now 4 times as many stamps as her. Had Oliver lost 44 stamps to Jill, he would have had 1½ times as many stamps as her. How many stamps did Oliver have at first?

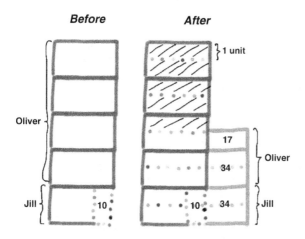

From the model,

5 units → 34 + 34 + 17 = 85
1 unit → 85 ÷ 5 = 17
8 units → 8 × 17 = 136
8 units + 10 → 136 + 10 = 146

Note: The total number of stamps remains unchanged.

Oliver had 146 stamps at first.

Check:
Oliver: 146; Jill: 2 × 17 − 10 = 24

	Before	After
Oliver	146 − 10 = 136	146 − 44 = 102
Jill	24 + 10 = 34	24 + 44 = 68
Ratio	136 : 34 = 4 : 1	102 : 68 = 3 : 2

Worked Example 46

Ken and Mary have some money each. If Ken gives Mary $16, both will have the same amount of money. If Mary gives Ken $9, Ken will have six times as much money as Mary. How much do Ken and Mary have in all?

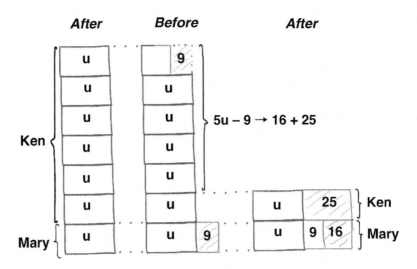

From the stack model,

$5u - 9 \rightarrow 16 + 25$
$5u \rightarrow 16 + 25 + 9 = 50$
$u \rightarrow 50 \div 5 = 10$
$7u \rightarrow 7 \times 10 = 70$

Ken and Mary have $70 in all.

Thought Process

Ken and Mary have some money each. If Ken gives Mary $16, both will have the same amount of money. If Mary gives Ken $9, Ken will have six times as much money as Mary. How much do Ken and Mary have in all?

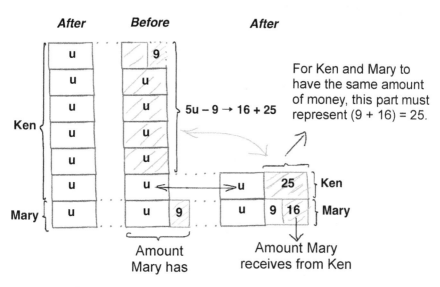

For Ken and Mary to have the same amount of money, this part must represent (9 + 16) = 25.

Amount Mary has

Amount Mary receives from Ken

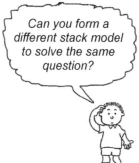

Can you form a different stack model to solve the same question?

From the stack model,

$5u - 9 \rightarrow 16 + 25$
$5u \rightarrow 16 + 25 + 9 = 50$
$u \rightarrow 50 \div 5 = 10$
$7u \rightarrow 7 \times 10 = 70$

Ken and Mary have $70 in all.

Worked Example 47

Ronny and Jeff have some stamps each. If Ronny were to give Jeff 12 stamps, both would have the same number of stamps. However, if Ronny were to give Jeff 40 stamps, Jeff would have three times as many stamps as Ronny. How many stamps does Jeff have?

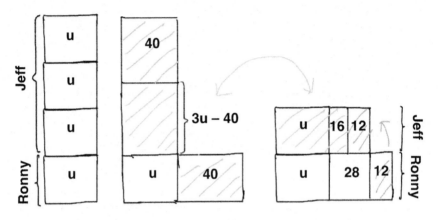

From the stack model,

$3u - 40 \rightarrow u + 16$
$u + 2u - 40 \rightarrow u + 16$
$2u - 40 \rightarrow 16$
$2u \rightarrow 16 + 40 = 56$
$u \rightarrow 56 \div 2 = 28$
$3u - 40 \rightarrow 3 \times 28 - 40 = 44$

Jeff has 44 stamps.

Check

	Jeff	Ronny
Before	44	68
	+ 12	− 12
After	56	56

	Jeff	Ronny
Before	44	68
	+ 40	− 40
After	84	28

$84 = 3 \times 28$

Thought Process

Ronny and Jeff have some stamps each. If Ronny were to give Jeff 12 stamps, both would have the same number of stamps. However, if Ronny were to give Jeff 40 stamps, Jeff would have three times as many stamps as Ronny. How many stamps does Jeff have?

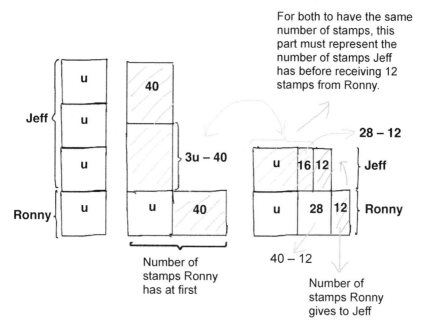

From the stack model,

$3u - 40 \rightarrow u + 16$
$2u \rightarrow 16 + 40 = 56$
$u \rightarrow 56 \div 2 = 28$
$3u - 40 \rightarrow 3 \times 28 - 40 = 44$

Jeff has 44 stamps.

Worked Example 48

When 30 cm³ of water from jug Q is poured into jug P, the volume of water in jug Q is 4 times the volume of water in jug P. When 50 cm³ of water from jug Q is poured into jug P, jug Q has 3 times as much water as jug P. What is the volume of water in jug Q?

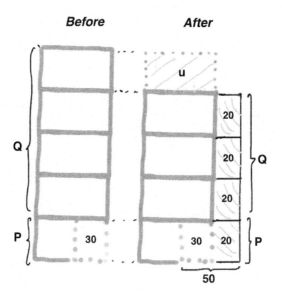

Since the total volume of water in jugs P and Q is unchanged in both cases,

u → 4 × 20 = 80

Volume of water in jug Q = 4 × 80 + 30 = 350 cm³

Thought Process

When 30 cm³ of water from jug Q is poured into jug P, the volume of water in jug Q is 4 times the volume of water in jug P. When 50 cm³ of water from jug Q is poured into jug P, jug Q has 3 times as much water as jug P. What is the volume of water in jug Q?

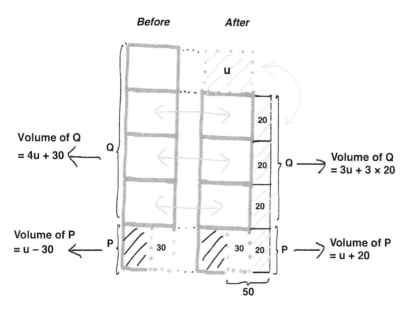

Since the total volume of water in jugs P and Q is unchanged in both cases,

u → 4 × 20 = 80

Volume of water in jug Q = 4u + 30 = 4 × 80 + 30 = 350 cm³

Something to think about ...
The Stack Method

Bar modeling is often presented like dozens of lines of software code. We need a simple model that requires half a dozen lines, which is easier to debug or alter. Stack modeling may be what we need to unclutter our otherwise oft-visually constipated minds!

The eye sees the colorful model drawings; the mind's eye sees the relationship between the bars, stacks, and dotted lines.

Bar or stack modeling is like pre-algebraic problem solving.

Is stack modeling a disruptive methodology?

Stack modeling is to disruptive methodology what NEWater is to disruptive technology.

Stack modeling is like creative visualization: you try to come up with the simplest model where the numerical relationship would pop up before your eyes.

Revision Questions

Here is a sample of **26 miscellaneous grades 5–6 word problems** that may be solved using the stack model method.

The aim here is to encourage the students to solve the questions the "Stack Model Way," then to compare their solutions against other problem-solving strategies, such as the bar model and Sakamoto methods. *Which types of word problems are favorable to the stack model method?*

0. In a class, 3/4 of the students are boys. 1/2 of the boys and 1/3 of the girls are on Facebook. What fraction of the class are on Facebook?

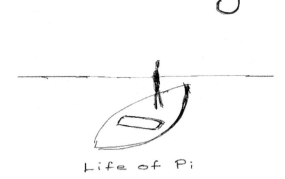

Life of Pi

Hint: Use a model to illustrate the different parts.

1. Mrs. Yan bought some candies for her students. If she gave each student 4 candies, she would have 2 candies left. If she gave each student 5 candies, she would need one more candy.
   (a) How many candies did she buy?
   (b) How many students were there?

Hint: See Worked Examples 3 and 4.

2. Conrad and Barry downloaded a total of 63 PowerPoint slides to their websites. After Conrad added 6 more slides and Barry added 27 more slides, Barry had now three times as many slides as Conrad. How many PowerPoint slides did each person download at first?

Try using two different stack models to solve this question. How does each fare as compared to the bar model?

3. Sheila had $250 and Doris had $75 at first. After Doris received $25 less than Sheila, Doris had one third as much left as Sheila. How much money did Doris receive?

Hint: See Worked Example 7.

4. Mrs. Jeff is in charge of 45 students. After selecting 1/7 of the boys and 6 girls to take part in a Science contest, she found that there is an equal of boys and girls who were not selected for the contest. How many girls are there in the class?

To Tweet or Not to Tweet

@MathPlus #stackmath

5. In 1985, Donald was three times as old as Jimmy. In 2003, Donald was twice as old as Jimmy. How old was Donald in 2010?

Age is just a number! You can be as young as you want to be!

Hint: See Worked Example 10.

6. When Ian was born, his mum was three times as old as Doris. When Ian was 25 years old, his mum was twice as old as Doris. How old was Doris when Ian was born?

7. A group of tourists visited a Toys museum. The admission tickets cost $9 per adult and $6 per child. They paid a total of $408 for admission tickets. If the ratio of the number of adults to the number of children was 8 : 5, how many adults were there in the group?

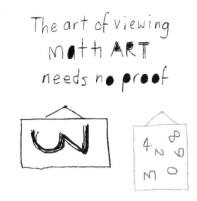

8. In a class, 2% of the students were absent. Of those present, 5/7 passed a Tamil test, and the rest failed. What percents of the whole class failed the test?

9. At a university, students who fail a math test sit for a re-test. 70% of the students passed in their first attempt, and 1/3 of those who took the re-test failed again. What percents of students passed the test or re-test?

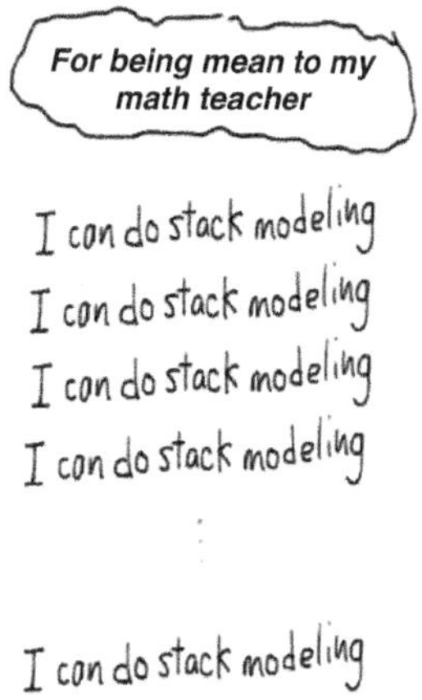

> For being mean to my math teacher

I can do stack modeling
I can do stack modeling
I can do stack modeling
I can do stack modeling
:
I can do stack modeling

10. The ratio of Vani's age to Tom's age is 3 : 1 now. In 25 years' time, the ratio will become 4 : 3. How old is Vani's now?

11. The number of grade 5 students is 25% less than the number of grade 4 students. The number of grade 3 students is 5% more than the number of grade 4 students. If there are 78 more grade 3 students than grade 5 students, what is the total number of grades 3–5 students?

> I Google, therefore I am.
>
> **Dr. Googol**

12. Carl, Dave, and Esther decided to help their friend Fiona. Carl, Dave, and Esther gave 1/4, 1/5, and 1/3 of their money, respectively, to Fiona. In the end, Fiona received the same amount of money from each of her friends. What percents of the total amount of money is owned by Fiona now?

13. Gina had 7/13 of the number of Facebook likes Gerald had. Six months later, after each had received the same number of new likes, Gina had now 9/10 of the number of likes Gerald had. Given that the number of new Facebook likes lies between 500 and 550, how many Facebook likes did Gina receive at first?

14. When Sam stands on a chair of height 0.35 m and Ann stands on a stool, Sam is 0.8 m taller than Ann. However, when both stand on the floor, Sam is 0.75 m taller than Ann. What is the height of the stool?

* The publisher refused to submit this book for approval —although the author was itching to do so in the hope of boosting exponential sales.

15. The total age of Dave and Esther is three times the age of Carol. Five years from now, the difference between Dave and Esther's total age and Carol's age will be 37 years. How young is Carol now?

Keep your mind young, no matter what your age is!

16. A company employs 320 men, three-quarters of whom are married. 3/4 of the married men have at least one child and 3/4 of these fathers have more than one child. How many have exactly one child?

Muslims don't eat pork

Hindus don't eat beef.

...

Mathematicians don't eat ─── .

17. One quarter of Facebook friends and two-fifths of Twitter followers sum up to 21. Three-quarters of Facebook friends and four-fifths of Twitter followers add up to 47. How many Facebook friends and Twitter followers are there altogether?

How can I solve this question *without* algebra?

18. Fanny and Dave had the same amount of money. After Fanny bought an e-book for $14 and Dave bought a graphing calculator for $110, Fanny had 5 times as much money as Dave. How much money did Fanny have at first?

> *Stack modeling* is for cats; *bar modeling* is for dogs.

Hint: See Worked Examples 6 and 7.

19. The ratio of the amount of money Henry had to the amount of money Lily had was 3 : 2. After Henry gave away $5 and Lily received $8, the ratio became 5 : 4. How much money did Lily have at first?

Where Mr. Yan wrote this e-book post-Christmas.

Hint: See Worked Examples 29, 30, and 32.

20. Ian and Sue have some money each. If Ian gives Sue $21, both will have the same amount of money. If Sue gives Ian $12, Ian will have four times as much money as Sue. How much do Ian and Sue have in all?

Stack
~~Bar~~ Modeling
is ~~a science.~~
an art.

The Whole Story
Hole

Hint: See Worked Examples 44–47

21. Sally and David have some marbles each. If Sally were to give David 13 marbles, both would have the same number of marbles. However, if Sally were to give David 37 marbles, Sally would have one-quarter as many marbles as David. How many marbles does David have?

Not all stack models are created equal — *some are more elegant than others*.

Hint: See Worked Examples 44–47

22. There were 65 people who logged in to a free webinar. When 3/4 of the men and 3/5 of the women logged off, the total number of men and women who remained online became 20. How many women logged in at first?

Hint: See Worked Example 20.

23. A tank with 171 liters of water is divided into three containers, A, B, and C. Container B has three times as much water as container A. Container C has 1/4 as much water as container B. How much water is there in container B?

The *good news* is, we now have this book.

24. When the Smiths got married 18 years ago, Mr. Smith was three times as old as his wife. Today, he is only twice as old as she is. How old was Mrs. Smith when she got married?

Singapore Jubilee
#SG50

25. The ratio of Andrew's money to Kevin's money was 4 : 3 at first. After Andrew got $120 and Kevin spent $240, the ratio of Andrew's money to Kevin's money became 13 : 7. How much does Andrew have now?

Answers & Solutions

Encourage students to solve each question in as many ways as possible, besides using the Stack Model Method. Tickle the students to come up with a different stack model, other than the one(s) given in this book.

Remember: *Not all stack or bar models are created equal —some are more elegant than others.*

0. In a class, 3/4 of the students are boys. 1/2 of the boys and 1/3 of the girls are on Facebook. What fraction of the class are on Facebook?

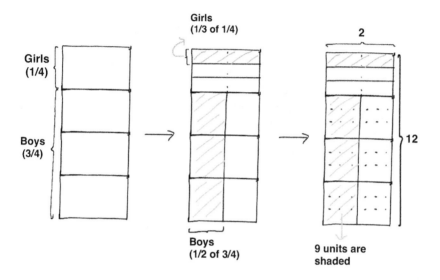

From the model, the entire class can be represented by (2 × 12) = 24 units.

Out of these 24 units, 11 units are shaded.

Fraction of the class which are on Facebook = 11/24

1. Mrs. Yan bought some candies for her students. If she gave each student 4 candies, she would have 2 candies left. If she gave each student 5 candies, she would need one more candy.
   (a) How many candies did she buy?
   (b) How many students were there?

Let one unit (☐) represent the number of students.

Since the total number of candies does not change, we can equate this number based on the number of candies given to each student as follows:

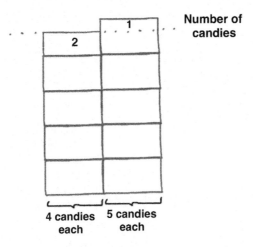

From the model,

1 unit → 2 + 1 = 3
So there are 3 students.

4 units + 2 → 4 × 3 + 2 = 12 + 2 = 14
or
5 units − 1 → 5 × 3 − 1 = 15 − 1 = 14

Mrs. Yan bought 14 candies.

2. Conrad and Barry downloaded a total of 63 PowerPoint slides to their websites. After Conrad added 6 more slides and Barry added 27 more slides, Barry had now three times as many slides as Conrad. How many PowerPoint slides did each person download at first?

Method 1

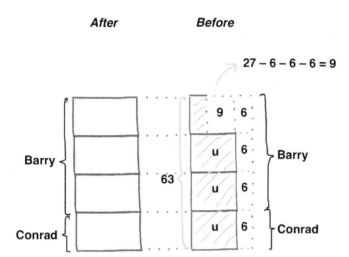

From the model,

4u → 63 + 9 = 72
u → 72 ÷ 4 = 18

Conrad downloaded 18 PowerPoint slides.

63 − 18 = 45
Barry downloaded 45 PowerPoint slides.

Conrad and Barry downloaded a total of 63 PowerPoint slides to their websites. After Conrad added 6 more slides and Barry added 27 more slides, Barry had now three times as many slides as Conrad. How many PowerPoint slides did each person download at first?

Method 2

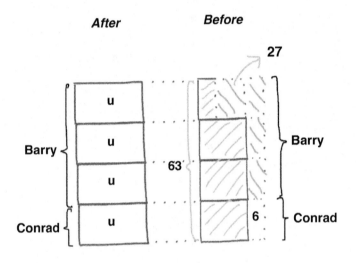

From the model,

4u → 63 + 6 + 27 = 96
u → 96 ÷ 4 = 24
u − 6 → 24 − 6 = 18
Conrad downloaded 18 PowerPoint slides.

63 − 18 = 45
Barry downloaded 45 PowerPoint slides.

3. Sheila had $250 and Doris had $75 at first. After Doris received $25 less than Sheila, Doris had one third as much left as Sheila. How much money did Doris receive?

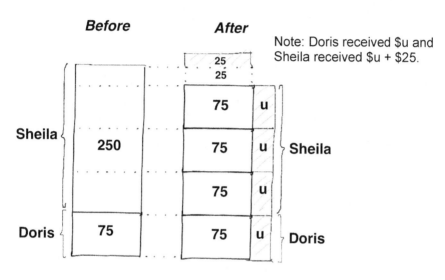

From the model,

2u → 25 + 25
u → 25

Doris received $25.

Thought Process

Sheila had $250 and Doris had $75 at first. After Doris received $25 less than Sheila, Doris had one third as much left as Sheila. How much money did Doris receive?

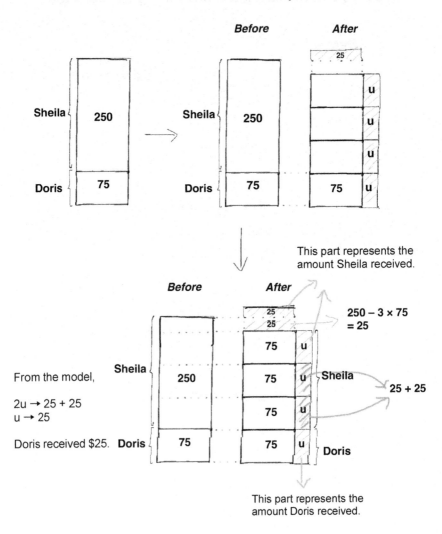

From the model,

2u → 25 + 25
u → 25

Doris received $25.

4. 24 girls.

Method 1

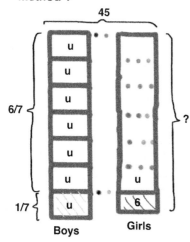

From the model,

7u + 6u → 45 − 6 = 39
13u → 39 ÷ 13 = 3
6u + 6 → 6 × 3 + 6 = 18 + 6 = 24

There are 24 girls in the class.

Method 2

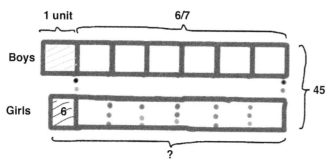

From the model,

7 units + 6 units → 45 − 6 = 39
13 units → 39 ÷ 13 = 3
6 units + 6 → 6 × 3 + 6 = 18 + 6 = 24

There are 24 girls in the class.

5. In 1985, Donald was three times as old as Jimmy. In 2003, Donald was twice as old as Jimmy. How old was Donald in 2010?

2003 − 1985 = 18

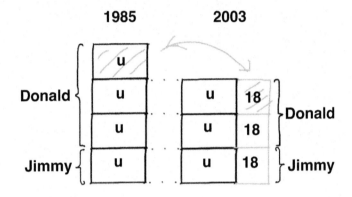

From the model,

u → 18
3u → 3 × 18 = 54
Donald was 54 years old in 1985.

2010 − 1985 = 25
54 + 25 = 79

In 2010, Donald was 79 years old.

6. 25 years old.

Given: When Ian was 25 years old, his mum was twice as old as Doris.

This means that 25 years later (after Ian was born), Ian's mum will be 2 times as old as Doris.

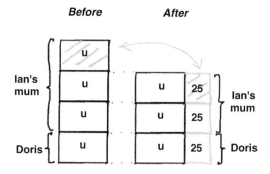

From the model,

u → 25

Doris was 25 years old when Ian was born.

7. 32 adults.

Adults   Children
  8   5

8 units × 9 + 5 units × 6 = 408
    72 units + 30 units = 408
          102 units = 408
              1 unit = 408 ÷ 102 = 4
            8 units = 8 × 4 = 32

There were 32 adults.

8. 28%.

Suppose there were 100 students in the class.

2% of the students were absent. This means that
2/100 × 100 = 2 students were absent.

5/7 passed the test.
1 − 5/7 = 2/7 failed the test.

Out of the remaining 100 − 2 = 98 students who
were present, 2/7 × 98 = 28 failed the test.

Now, 28/100 × 100% = 28%
28% of the whole class failed the test.

TH!NK

Is it mathematically correct to solve the problem as follows?

2/7 × (100 − 2) = 2/7 × 98 = 28
So, 28% of the class failed the test.

9. 90%.

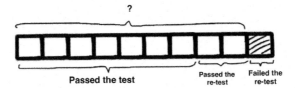

From the model,

9 out of 10 units represent students who passed the test or re-test.

9/10 × 100% = 90%

90% of students passed the test or retest.

10. 15 years old.

Method 1

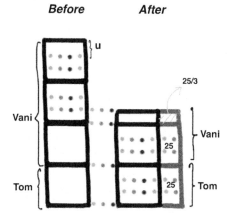

From the model,

5u → 25/3
9u → 25/3 × 9/5 = 15

Vani is 15 years old.

Check:

3u → 15 ÷ 3 = 5
Tom is 5 years old now.

Vani	Tom	Note: Using the bar-
15	5	instead of the stack-
+ 25	+ 25	model method to solve
------	------	this question may prove
40 : 30		quite challenging.
4 : 3		

171

10. The ratio of Vani's age to Tom's age is 3 : 1 now.
In 25 years' time, the ratio will become 4 : 3. How old is Vani's now?

Method 2

Using the units method

Before

Vani: 3 units
Tom: 1 unit

After
Vani: 4 units × 2 = 8 units
Tom: 3 units × 2 = 6 units

Note: We do some guess and check to the "before" and "after" cases until the differences in units in both cases between Vani and Tom are the same, which is 5 units.

Increase in units = 8 − 3 = 5 or 6 − 1 = 5
5 units → 25 years
1 unit → 25 ÷ 5 = 5 years
3 units → 3 × 5 = 15 years

Vani is 15 years old now.

Note: The "units method" may appear to be simpler (or neater) in presentation, but that doesn't mean that it takes a shorter time, because a few guesses and checks may be needed before the two differences are found to be equal.

11. 728.

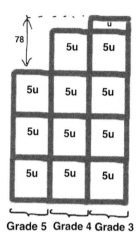

Grade 5 Grade 4 Grade 3

25% = 1/4; 5% = 5/100 = 1/20

From the model,
6u → 78
u → 78 ÷ 6 = 13
11 × 5u + u = 56u → 56 × 13 = 728

The total number of grades 3–5 students is 728.

12. 25%.

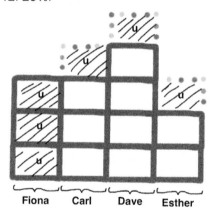

Fiona Carl Dave Esther

From the model,

Carl, Dave, and Esther owned 12 units of money.
Fiona now owns 3 units.

3/12 × 100% = 25%
Fiona owns 25% of the total amount of money.

13. 77 likes.

Method 1

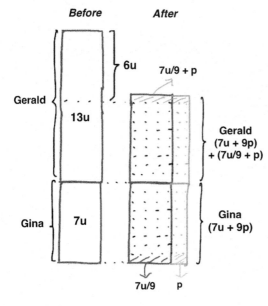

From the model,

$6u \rightarrow 7u/9 + p$

$p \rightarrow 6u - 7u/9 = 54u/9 - 7u/9 = 47u/9$

$9p \rightarrow 47u$

Now, $500 < 47u < 550$
Since u is a whole number, 47u must be 517, which occurs when $u = 11$.

Thus, $7u = 7 \times 11 = 77$

Gina received 77 Facebook likes at first.

Method 2

Using the units method

Before
Gina: 7 units
Gerald: 13 units

After
Gina: 9 units × 6 = 54 units
Gerald: 10 units × 6 = 60 units

Increase in units = 54 − 7 = 47 or 60 − 13 = 47
47 units must lie between 500 and 550.

Since 1 unit must be a whole number, 47 units can only be 517—when 1 unit = 11.

1 unit → 11 Facebook likes
7 units → 7 × 11 = 77 Facebook likes

Gina had 77 Facebook likes at first.

Note: We do some guess and check to both the "before" and "after" cases, until the difference in units between Gina and Gerald in either case is equal, which is 47 units.

14. 0.3 m.

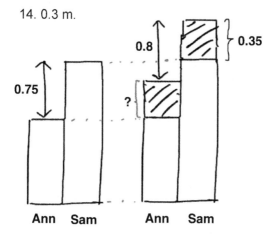

From the model,
height of the stool = 0.75 + 0.35 − 0.8 = 0.3 m

The height of the stool is 0.3 m.

15. 16 years old.

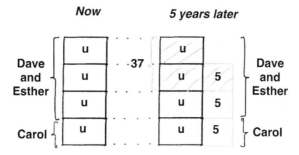

From the model,

2u + 5 → 37
2u → 37 − 5 = 32
u → 32 ÷ 2 = 16

Carol is 16 years old now.

16. 45.

3/4 × 320 = 240 of the men are married.

Out of the 240 married men, 3/4 × 240 = 180 of them have at least one child.

Out of the 180 of these fathers, 3/4 of them have more than one child. In other words, (1 − 3/4) = 1/4 of them have only one child.

1/4 × 180 = 45
So, only 45 of these fathers have exactly one child.

17. 60.

47 − 2 × 21 = 5

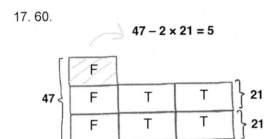

From the model,

F → 47 × 2 × 21 = 5
4F → 4 × 5 = 20
There are 20 Facebook friends.

F + 2T → 21 (given)
5 + 2T → 21
2T → 21 − 5 = 16
T → 16 ÷ 2 = 8
5T → 5 × 8 = 40
There are 40 Twitter followers.

Altogether, there are (20 + 40) = 60
Facebook friends and Twitter followers.

18. $134.

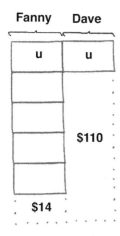

From the model,

4u → 110 − 14 = 96
u → 96 ÷ 4 = 24
5u → 5 × 24 = 10 × 12 = 120
5u + 14 → 120 + 14 = 134

Fanny had $134 at first.

19. $60.

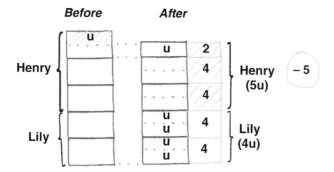

From the model,

u → 2 + 4 + 4 + 5 = 15
4u → 4 × 15 = 60

Lily had $60 at first.

Check

Before
Henry: 6u → 6 × 15 = 90
Lily: 60
Henry : Lily = 90 : 60 = 3 : 2

After
Henry: 90 − 5 = 85 = 5 × 17
Lily: 60 + 8 = 68 = 4 × 17
Henry : Lily = 5 : 4

20. $110.

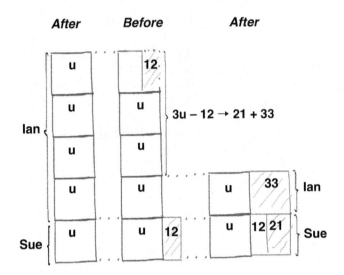

From the stack model,

3u − 12 → 21 + 33
3u → 21 + 33 + 12 = 66
u → 66 ÷ 3 = 22
5u → 5 × 22 = 110

Ian and Sue have $110 in all.

Check

	Before	After
Sue	u + 12 = 34	34 − 12 = 22
Ian	4u − 12 = 76	76 + 12 = 88
Sue		34 + 21 = 55
Ian		76 − 21 = 55

21. 27 marbles.

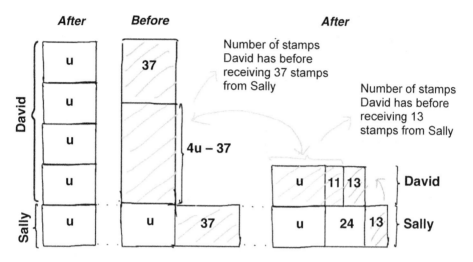

From the stack model,

$4u - 37 \rightarrow u + 11$
$u + 3u - 37 \rightarrow u + 11$
$3u - 37 \rightarrow 11$
$3u \rightarrow 11 + 37 = 48$
$u \rightarrow 48 \div 3 = 16$
$4u - 37 \rightarrow 4 \times 16 - 37 = 27$

David has 27 marbles.

Check

	David	Sally
Before	27	53
	+ 13	− 13
After	40	40

	David	Sally
Before	27	53
	+ 37	− 37
After	64	16

$64 = 4 \times 16$

22. 25 women.

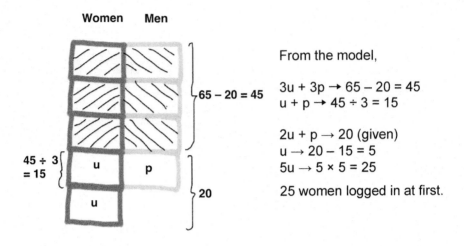

From the model,

3u + 3p → 65 − 20 = 45
u + p → 45 ÷ 3 = 15

2u + p → 20 (given)
u → 20 − 15 = 5
5u → 5 × 5 = 25

25 women logged in at first.

23. 108 liters.

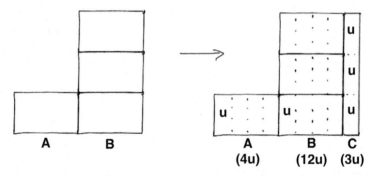

From the model,

19u → 171
u → 171 ÷ 19 = 9
12u → 12 × 9 = 108

There were 108 liters of water in container B.

24. 18 years old.

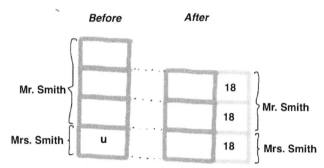

From the model,

Mr. Smith: 3u (*Before* → 2u + 18 (*After*)
u → 18

Mrs. Smith was 18 years old when she got married.

Check:
Before: Mr. Smith → 3 × 18 = 54; Mrs. Smith → 18
After: Mr. Smith → 54 + 18 = 72; Mrs. Smith → 18 + 18 = 36
72 = 2 × 36

Thought Process

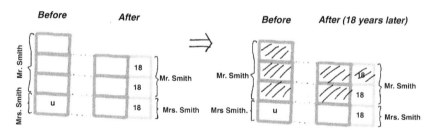

Mr. Smith: 3u (*Before*) → 2u + 18 (*After*)

25. $1560.

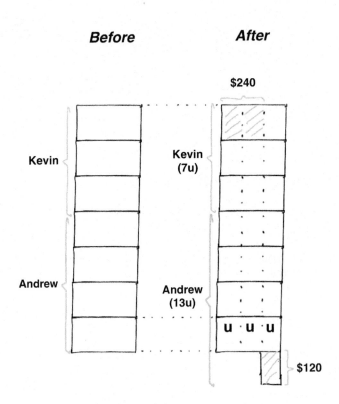

From the model,

u → $120

13u → 13 × $120 = $1560

Andrew has now $1560.

Bibliography and References

Tan, S. (2010). *Model approach to problem-solving: Stack & split to solve challenging problems fast!* Singapore: Maths Heuristics (S) Private Limited.

Tan, S. Y. (undated). *Mathswise strategies book upper primary.* Singapore: Horizon Distribution Centre.

Wan, C. H. (2006). *Challenging maths problems made easy.* Singapore: Marshall Cavendish Education.

Yan, K. C. (2015). *Word problems in focus—Primary 6A.* Singapore: Marshall Cavendish Education.

Yan, K. C. (2015). *Word problems in focus—Primary 6B.* Singapore: Marshall Cavendish Education.

Yan, K. C. (2015). *Word problems in focus—Primary 5A.* Singapore: Marshall Cavendish Education.

Yan, K. C. (2015). *Word problems in focus—Primary 5B.* Singapore: Marshall Cavendish Education.

Yan, K. C. (2014). *Primary mathematics challenging word problems (Common Core Edition)—Grade 5.* Singapore: Marshall Cavendish Education.

Yan, K. C. (2014). *Primary mathematics challenging word problems (Common Core Edition)—Grade 4.* Singapore: Marshall Cavendish Education.

About the Author

Kow-Cheong Yan is the author of Singapore's best-selling *Mathematical Quickies & Trickies* series and the coauthor of the MOE-approved *Additional Maths 360*. Besides coaching mathletes and conducting recreational math courses for students, teachers, and parents, he edits, ghostwrites, and consults for *MathPlus Consultancy*.

Kow-Cheong writes about the good, the bad, and the ugly of Singapore's math education and of the local educational publishing industry. Read his two math blogs at **www.singaporemathplus.com** and **www.singaporemathplus.net**, or follow him on Twitter @MathPlus, @Zero_Math, and @SakamotoMath.

His e-mail coordinates are
kcyan@singaporemathplus.com and
kcyan.mathplus@gmail.com.

Visit Kow Cheong's Facebook Pages and Pinterest Boards at these virtual addresses:

fb.com/SingaporeMathPlus
fb.com/AddMaths360
fb.com/Christmaths
pinterest.com/MathPlus